Embedded Linux Development Using Eclipse

Embedded Linux Development Using Eclipse

Doug Abbott

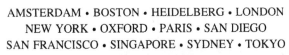

AMSTERDAM • BOSTON • HEIDELBERG • LONDON
NEW YORK • OXFORD • PARIS • SAN DIEGO
SAN FRANCISCO • SINGAPORE • SYDNEY • TOKYO

Newnes is an imprint of Elsevier

Newnes is an imprint of Elsevier
30 Corporate Drive, Suite 400, Burlington, MA 01803, USA
Linacre House, Jordan Hill, Oxford OX2 8DP, UK

 Recognizing the importance of preserving what has been written, Elsevier prints its books on acid-free paper
whenever possible.

Library of Congress Cataloging-in-Publication Data
Abbott, Doug, 1944-
 Embedded Linux development using Eclipse/Doug Abbott.
   p. cm.
 Includes index.
 ISBN 978-0-7506-8654-9
 1. Linux. 2. Eclipse (Electronic resource) 3. Embedded computer systems–Programming.   I. Title.
 QA76.76.O63A237 2008
 005.4′32–dc22

                                                                                                2008041350

British Library Cataloguing-in-Publication Data
A catalogue record for this book is available from the British Library.

ISBN: 978-0-7506-8654-9

For information on all Newnes publications,
visit our Web site at: www.books.elsevier.com

Transferred to Digital Printing, 2010

Printed and bound in the United Kingdom

On a personal level, this book is dedicated to the two most
important people in my life:

To **Susan,** my best friend, my soul mate. Thanks for sharing life's journey with me.

To **Brian**, budding lighting designer, actor and musician, future pilot, and all-around
great kid. Thanks for keeping me young at heart.

*Never doubt that a small band of committed citizens can change the world. Indeed,
it is the only thing that ever has.*

—Margaret Mead

On a professional level, the book is dedicated to those who made it possible:
open source programmers everywhere, especially that small band of open source
pioneers whose radical notion of free software for the good of the community,
and just for the fun of it, has indeed changed the world.

Table of Contents

Foreword: A Brave New World of Embedded Software Development

The world of embedded computing has always struggled to offer its developers a consistent user interface. In the early days of 8-bit processors, the tools required to program them were very specific to the processor architecture, and developers used in-circuit emulators to debug their machine-code based applications. Then, as high level languages emerged, a consistency and portability of embedded code emerged. Sadly, the user interface to use and debug these high level applications was typically proprietary to the tools vendor providing them. The open-source community provided good compiler tools such as GNU GCC, but the debug interface for GNU GDB was basic, and had no concept of an Integrated Development Environment (IDE). The closest any embedded vendors got to a consistent user interface was to use Microsoft's Visual Studio extendibility to produce an embedded IDE, which was both beholden to Microsoft and only available on a Windows platform.

Fast forward to around 2003, and some embedded vendors (typically providing embedded Linux or similar OS solutions) decided to investigate using an open-source framework called Eclipse. This framework was not designed to be an embedded IDE, but was open and extendible enough that it could be made to work. These early embedded vendors also got involved with the Eclipse consortium and then the Eclipse foundation, and helped steer both the framework and some of the most relevant Eclipse projects to meet the needs of embedded developers.

Today in 2008, the Eclipse framework is used by the majority of embedded tools and operating system providers, and finally offers a consistent user interface for embedded developers regardless of which processor, operating system, host development platform or language is being used. In a matter of four years Eclipse has gone from relative obscurity to the de-facto environment for embedded software development. Embedded Systems Conferences will often have an Eclipse track, and even the EclipseCon conference typically has embedded tutorials, showing that Eclipse is important to embedded, and embedded is important to Eclipse.

Doug's book is an important step for the embedded community as it guides the reader through the important first steps of using Eclipse for embedded development, especially when using an embedded Linux operating system. Although Eclipse is generally intuitive to a novice user, it is somewhat of a paradigm shift for embedded developers who have typically had a very "debugger-centric" view of the world. In Eclipse, the code is the center, and a debugger is just one of a selection of tools that are available to work with the code.

By reading this book from start to finish, the embedded developer will gain a rare insight into everything Eclipse: from its history, through a guided installation of Eclipse and CDT, and then into examples of new features, new projects, plug-ins, and commercial offerings. This will give the embedded developer the confidence and understanding to become very productive very quickly when using this new environment.

The many months that Doug has spent researching and evaluating Eclipse and then carefully detailing his findings in this book will serve our embedded community well, and will help to further propagate the adoption of this unique standard platform for embedded Linux developers across the globe.

Robert Day
VP Marketing, Lynux Works
Solutions Members Representative,
Board of Directors,
Eclipse Foundation Chairman,
Embedded Workgroup,
Eclipse Foundation

Preface

The open source software movement has come a long way since Richard Stallman founded the Free Software Foundation in 1985. Once the province of hackers and hobbyists, open source has gone "mainstream." Linux is now found in cell phones, PDAs, and countless other embedded devices. It's even starting to make inroads on the desktop. Apache is probably the most widely used web server now. While the business models are not always clear, numerous companies are finding a niche commercializing open source in one form or another.

Then there's Eclipse. Who would have thought, say 10 years ago, that the movers and shakers of the software industry, the likes of Borland, IBM, and Nokia, would come together and actually *cooperate* on a major open source platform for software development?

But we should all be very grateful they did. The result is, in my opinion, probably the most professional, well-managed open source project around. In terms of quality, ease of use, and just plain "polish," Eclipse is as good as, if not better, than any other IDE I've used in my long career as a software developer. The breadth of functionality offered by the Eclipse ecosystem is truly astounding, encompassing Java and web development, enterprise development and business tools, and embedded device software development, to name just a few.

Embedded software development using Eclipse is the subject of this book. Many books have been written about Eclipse, but none have yet addressed this particular niche. This seemed like a good time. The C Development Toolkit (CDT) and Device Software Development Platform (DSDP) in particular seem to have made great strides with the June 2008 Ganymede release of Eclipse.

Specifically, the book focuses on embedded software development, using Linux as both the host workstation and as the target. Why Linux? Several reasons, really:

- It's free, and so are the tools needed to build software.

- Linux derives from the same open source impulse that spawned Eclipse.

- I happen to like Linux, although there are times when I wonder why.

Audience and Prerequisites

While the primary focus is embedded software development, much of the book is applicable to anyone developing software in C or C++. Thus, it is expected that you have a reasonably good understanding of C and/or C++, know how to build executable applications from source code, and how to debug and run those applications—if not on an embedded target, at least on a workstation. You should at least know what a makefile is even if you're not fluent in the details of the make language.

The required Linux background is fairly minimal. You should know your way around the file system and how to change file permissions, how to install software from archive files, and a little bit about scripting. Unlike most Linux hackers, I do most of my work from the KDE graphical desktop environment and only use the command shell when absolutely necessary. That's a personal preference, and of course, you're free to adopt whatever working style suits you.

No previous experience with Eclipse itself is assumed. If you've worked with other IDEs, so much the better.

This is very much a "hands-on" book. To get the most out of it, you need to actually do the steps that are being described. Of course, Eclipse runs equally well on Windows and Mac OS platforms. It's just that it's harder to set up a software development environment on those platforms. If you choose to use a Windows platform, Chapter 2, "Installation," offers some guidance on setting up an appropriate Windows environment.

Resources

http://www.intellimetrix.us—Here's where you can purchase the embedded target board mentioned in Chapter 6 and described in more detail in Appendix B. The downloads page will host any updates or corrections.

http://www.groups.yahoo.com/group/eclipsebook—This Yahoo! group is specifically for readers of the book, a place to ask questions and share ideas. Updates and corrections will also be posted here.

Acknowledgments

This project never would have come to fruition without my editor, Rachel Roumeliotis, and her assistant, Heather Scherer. They had the thankless job of continually "prodding" me to stay on schedule and get the thing done!

Several Eclipse contributors on the newsgroups pointed me in the right direction when I was puzzling over some seemingly strange behavior: Wayne Beaton, Doug Gaff, David McKnight, Shigeki Moride, and Martin Oberhuber.

Thanks are also due to Craig Crinklaw and Sonia Leal at LynuxWorks, and Troy Kitch at Monta Vista for supplying me with, and helping me get running, demos of their Eclipse-based IDE products.

CHAPTER 1

Introducing Eclipse

Stroll around the exhibit floor at the Embedded Systems Conference either in Silicon Valley in April, or Boston in September. You're sure to see any number of large, flashy booths with big flat screen monitors showing off the vendor's integrated software development environment. You'll probably also hear a pitch by a well-groomed marketing type wearing a wireless headset and offering a T-shirt or other giveaway if you'll hang around for the full presentation.

After a while, it begins to dawn on you that all these integrated development environments (IDEs) seem to have a similar look and feel. Coincidence? Not really. Turns out that most of the major embedded software vendors have adopted Eclipse as the foundation for their IDE products.

It makes perfect sense. Developing embedded development tools is expensive and time-consuming. Making use of a common platform saves a lot of that time and expense. It also provides tools with a consistent look and feel that run in a wide range of operating environments. Vendors compete on the basis of their own proprietary additions to the base platform.

And really, is it any different from the multitude of vendors who build embedded products based on the PC architecture? They all start with the same PC platform and then add their own proprietary hardware and software to create value-added products.

1.1 History

Eclipse grew out of the project that began in 1998 at Object Technology International (OTI), a subsidiary of IBM now known as the IBM Ottawa Lab. The project was initiated to address complaints raised by IBM's customers that the company's tools didn't work well together.

In 2001, IBM established the Eclipse consortium and released the entire code base, estimated to be worth $40 million at the time, as open source. The idea was to let the open source community control the code and let the consortium deal with commercial relations. The initial nine members of the consortium included both partners and competitors of IBM at the time, such as Rational and TogetherSoft.

As Eclipse grew and evolved, IBM wanted more serious commitment from vendors, but vendors were reluctant to make a strategic commitment as long as they perceived that IBM was in control. This problem was addressed in 2004 with the creation of the Eclipse Foundation, a not-for-profit organization with a professional staff and a large and growing roster of commercial software vendors as members.

As of August 2007, the Eclipse Foundation listed 166 members on its web site. There are four categories of membership reflecting different levels of commitment[1]:

- *Committers.* Individuals who contribute and commit code to Eclipse projects. They may be members by virtue of working for a member organization, or may choose to join independently if they are not. Note that Committers are not included in the membership total above, as they are not listed on the web site.

- *Associates.* These are standards organizations, research and academic institutions, open source advocates, or publishing houses that participate in the development of the Eclipse ecosystem. There are currently 24 associates who can submit requirements, participate in all project reviews, and participate fully in all Membership Meetings. Associate Members are not assessed dues.

- *Add-in Providers.* Commercial software vendors who have publicly expressed support for Eclipse. Add-in Providers are expected to make available a commercial Eclipse-based product or service within 12 months of joining the foundation. Products may be built using Eclipse tools or on top of Eclipse projects. Services may include, for example, training, consulting, or a hosted web service. The 121 Add-in Provider Members each pay annual dues of $5,000.

---

[1] The membership categories were modified somewhat in July of 2008. For details of the new membership categories, see the following page at the Eclipse Web site: www.eclipse.org/membership/become_a_member/membershipTypes.php.

- ***Strategic Members.*** These are the big dogs. Strategic Members fall into two categories: Strategic Developers and Strategic Consumers. Each Strategic Developer is expected to have at least eight developers assigned full time to developing Eclipse technology and contribute annual dues of 0.12% of revenue with no minimum and a maximum of $250K.

  Strategic Consumers are users of Eclipse technology. They contribute annual dues of 0.2% of revenue with a minimum of $50K and a maximum of $500K, but can reduce the cash outlay by contributing one or two developers to Eclipse projects at a rate of $125K for each developer.

  Each Strategic Member has a representative on the Eclipse Foundation Board of Directors allowing them direct control over the strategic direction of Eclipse. Strategic Members also have a seat on the Eclipse Requirements Council providing input and influence over the themes and priorities of Eclipse technology.

  There are currently 21 Strategic Members, including IBM, Intel, Motorola, Nokia, Oracle, Sybase, and Wind River, among others.

1.2 Eclipse Public License

Open source software is released to the public under the terms of a license that grants users of the software certain rights, the most significant of which is access to the source code. The software is copyrighted by its author, but rather than using the copyright to restrict access and use, which is the usual case, the copyright becomes the means of enforcing the rights granted by the open source license. Because open source effectively "reverses" the rights granted by copyright, it is often referred to as "copyleft."

Many users familiar with open source software may assume that the Gnu General Public License (GPL) is the one and only mechanism for making open source code available. In fact, there are over 50 different licenses certified by the Open Source Initiative, a non-profit body that reviews and certifies licenses that meet its 10-point definition of what constitutes open source software. Note, incidentally, that open source does not prohibit one from charging a fee to distribute open source software, and indeed, many companies are in the business of doing exactly that.

Part of the FUD (fear, uncertainty, and doubt) spread about open source software, mostly by a certain software company in the northwest United States, revolves around its so-called "viral" nature. The implication is that if you use any open source software in your product, it "infects" the rest of the code, forcing it all to be open source. This is at least partially true of software released under the GPL and it's an issue that developers must take into consideration if they wish to keep their own code proprietary. The GPL requires that any "derivative work," that is, code derived from GPL code, must also be released under the terms of the GPL.

The motivation then for other open source licenses is to encourage and support commercial use and distribution of open source software by allowing developers to maintain their own contributions as proprietary. Among the licenses that do this is the Eclipse Public License (EPL). Specifically, the EPL allows a developer to license his own contributions under the license of his choice provided its provisions don't conflict with the EPL.

This works well for software like Eclipse that is based on a "plug-in" concept to extend the base platform. Plug-ins are independent software modules that communicate with the platform through well-defined interfaces. So while the platform itself is open source, plug-ins may be proprietary.

1.3   Status of Eclipse

In addition to the Eclipse platform itself, Eclipse comprises dozens of tool-oriented and application-oriented projects that operate as independent open source projects. For the past three years, usually in June, the foundation has organized a coordinated major release of the platform along with a large number of the constituent sub-projects. This allows users to try out new features without worrying about version incompatibility among the various tools.

Interestingly, these releases are named after the moons of Jupiter. The 2008 release is named "Ganymede" and included 23 projects representing over 18 million lines of code.

The Eclipse sub-projects are grouped into 10 major project areas that include:

- The Eclipse Platform
- Eclipse Technology
- Business Intelligence and Reporting Tools
- Data Tools Platform

- Device Software Development Platform (DSDP)

- Modeling

- Service Oriented Architecture

- Development Tools

- Test and Performance Tools Platform

- Eclipse Web Tools Platform

Of particular interest to embedded developers are the Device Software Development Platform (DSDP) and the Development Tools projects. Under Development Tools is the C/C++ Development Tool (CDT) project, which is a major focus of this book and the basis for commercial IDEs using Eclipse.

1.4   So What Is Eclipse, Anyway?

Eclipse itself is not an Integrated Development Environment (IDE). Rather it is a collection of frameworks and tools for building IDEs and complex "rich-client" applications. The Eclipse Foundation's website describes it as "an extensible development platform, runtimes and application frameworks for building, deploying and managing software across the entire software lifecycle." One early technical overview paper described it thus: "The Eclipse Platform is an IDE for anything, and for nothing in particular."

Although Eclipse has a lot of built-in functionality, most of that functionality is very generic. It takes additional tools to extend the platform to work with new content types, to do new things with existing content types, and to focus the generic functionality on a specific task.

Eclipse is largely written in Java, and was originally developed for it. Consequently, it runs on any machine with a Java Runtime Environment (JRE). Figure 1.1 shows the platform's major components and APIs. The platform's principal role is to provide tool developers with mechanisms to use, and rules to follow, for creating seamlessly integrated tools. These mechanisms are exposed via well defined API interfaces, classes, and methods. The platform also provides useful building blocks and frameworks that facilitate developing new tools.

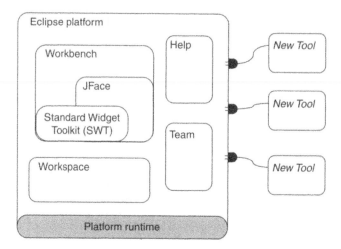

Figure 1.1: Elements of Eclipse.

Eclipse is designed and built to meet the following requirements:

- Support the construction of a variety of tools for application development.

- Support an unrestricted set of tool providers, including independent software vendors (ISVs).

- Support tools to manipulate arbitrary content types such as HTML, Java, C, JSP, EJB, XML, and GIF.

- Facilitate seamless integration of tools within and across different content types and tool providers.

- Support both GUI and non-GUI-based application development environments.

- Run on a wide range of operating systems, including Windows and Linux.

- Capitalize on the popularity of the Java programming language for writing tools.

1.4.1   Workbench

The *workbench* is the primary user interface for Eclipse. As such, it implements the Eclipse "personality" and supplies the structures that allow tools to interact with the user. Because of this central and defining role, the workbench is synonymous with

the Eclipse Platform UI as a whole and with the main window you see when
Eclipse is running.

The workbench, in turn, is implemented on top of two generic toolkits:

- *Standard Widget Toolkit (SWT).* A widget set and graphics library integrated
  with the native window system, but with an OS-independent API.

- *JFace.* A UI toolkit implemented using SWT that simplifies common UI
  programming tasks.

SWT provides a common API that works across a number of supported windowing
systems. For each native windowing system, SWT translates its common API into
native window widgets. Most common low-level widgets such as lists, text fields, and
buttons are implemented natively. But some generally useful higher-level widgets,
such as toolbars and trees, may need to be emulated on some window systems. The end
result is that SWT maintains a consistent programming model in all environments,
while preserving the look and feel of the underlying native window system. Thus,
Eclipse on a Mac looks like a Mac OS application, Eclipse under Windows XP (or Vista
if you prefer) looks like a Windows application, and so on.

JFace is a UI toolkit providing classes for handling many common UI programming
tasks such as image and font registries, dialog, preference, and wizard frameworks, and
progress reporting for long running operations. It sits on top of SWT and thus is
independent of the native windowing system.

1.4.2   Workspaces

The various tools plugged in to the Eclipse Platform operate on regular files in your
workspace. The workspace consists of one or more top-level *projects*, where each
project maps to a corresponding directory in the file system. The different projects in a
workspace may map to different file system directories or drives, although by default,
all projects map to sibling subdirectories of a single workspace directory.

It is also possible to have multiple workspaces. You specify a workspace when starting
Eclipse. From within Eclipse you can also change workspaces, which causes Eclipse
to restart itself.

A project contains files that you create and modify. In addition to being accessible from
Eclipse, all files in the workspace are directly accessible to the standard programs

and tools provided by the underlying operating system. Tools integrated with the Platform are provided with APIs for dealing with workspace *resources* (the collective term for projects, files, and folders). So-called *adaptable objects* represent workspace resources so that other parties can extend their behavior.

In a large project, the Linux kernel, for example, tools like compilers and link checkers must apply a coordinated analysis and transformation of thousands of separate files. To this end the platform provides an *incremental project builder* framework; the input to an incremental build is a *resource tree delta* capturing the net resource differences since the last build. The platform allows several different incremental project builders to be registered on the same project and provides ways to trigger project and workspace-wide builds. An optional workspace auto-build feature automatically triggers the necessary builds after each resource modification operation (or batch of operations).

1.4.3   Team Support

Eclipse supports programming teams with facilities for placing projects under the control of version and configuration management tools known as "team repository products." The Platform has extension points and a repository provider API that allow new kinds of team repositories to be plugged in.

Team repository products invariably affect the user's workflow, for example, by adding overt steps for retrieving files from the repository, for returning updated files to the repository, and for comparing different file versions. Eclipse allows each team repository provider to define its own workflow so that users already familiar with the native tool can quickly learn to use it from within Eclipse. The platform supplies basic hooks to allow a team repository provider to intervene in certain operations that manipulate resources in a project.

At the UI level, the platform supplies placeholders for certain actions, preferences, and properties, but leaves it to each repository provider to define these UI elements. There is also a simple, extendable configuration wizard that lets users associate projects with repositories, and which permits repository providers to extend with UI elements for collecting information specific to that particular repository.

Multiple team repository products can coexist peacefully within Eclipse. The platform includes built-in support for CVS repositories accessed via pserver, ssh, or extssh protocols.

1.4.4 Help

The Eclipse Platform Help mechanism allows tools (plug-ins) to define and contribute documentation to one or more online books. For example, a tool usually contributes help style documentation to a user's guide, and API documentation (if any) to a separate programmer's guide.

Raw content is provided as HTML files. The facilities for arranging the raw content into online books with suitable navigation structures are expressed separately in XML files. This separation allows pre-existing HTML documentation to be incorporated directly into online books without the need to edit or rewrite.

The add-on navigation structure presents the content of the books as a tree of topics. Each topic, including non-leaf topics, can have a link to a raw content page. A single book may have multiple alternate lists of top-level topics allowing some or all of the same information to be presented in completely different organizations. They may be organized by task, or by tool, for example.

The XML navigation files and HTML content files are stored in a plug-in's root directory or subdirectories. Small tools usually put their help documentation in the same plug-in as the code. Large tools often have separate help plug-ins. The Platform uses its own internal documentation server to provide the actual web pages from within the document web. This custom server allows the Platform to resolve special inter-plug-in links and to extract HTML pages from ZIP archives.

1.4.5 Plug-Ins

A *plug-in* is the smallest unit of Eclipse functionality that can be developed and delivered separately. A small tool is usually written as a single plug-in, whereas a complex tool may have its functionality split across several plug-ins. Except for a small kernel known as the Platform Runtime, all of the Eclipse platform's functionality as described above is located in plug-ins.

Plug-ins are coded in Java. A typical plug-in consists of Java code in a JAR library, some read-only files, and other resources such as images, web templates, message catalogs, native code libraries, etc. Some plug-ins don't contain code at all. An example is a plug-in that contributes online help in the form of HTML pages. A single plug-in's code libraries and read-only content are located together in a directory in the file system or at a base URL on a server.

Each plug-in's configuration is described by a pair of files. The *manifest* file, `manifest.mf`, declares essential information about the plug-in to other plug-ins, including the name, version, and dependencies. The second optional file, `plugin.xml`, declares the plug-in's interconnections to other plug-ins. The interconnection model is simple: a plug-in declares any number of named *extension points*, and any number of *extensions* to one or more extension points in other plug-ins.

The extension points can be extended by other plug-ins. For example, the workbench plug-in declares an extension point for user preferences. Any plug-in can contribute its own user preferences by defining extensions to this extension point.

On start-up, the Eclipse runtime discovers the set of available plug-ins, reads their manifest files, and builds an in-memory plug-in registry. The platform matches extension declarations by name to their corresponding extension point declarations. Any problems, such as extensions to missing extension points, are detected and logged. The resulting plug-in registry is available via the Platform API. After startup, plug-ins can be unloaded, and new ones installed or new versions of existing plug-ins can replace existing versions.

By default, a plug-in is *activated* when its code actually needs to be executed. Once activated, a plug-in uses the plug-in registry to discover and access the extensions contributed to its extension points. For example, the plug-in declaring the user preference extension point can discover all contributed user preferences and access their display names to construct a preference dialog. This can be done using only the information from the registry, without having to activate any of the contributing plug-ins. The contributing plug-in will be activated when you select one of its preferences from a list.

By determining the set of available plug-ins up front, and by supporting a significant exchange of information between plug-ins without having to activate any of them, the platform can provide each plug-in with a rich source of pertinent information about the context in which it is operating. The context doesn't change while the platform is running, so there's no need for complex life cycle events to inform plug-ins when the context changes. This avoids a lengthy start-up sequence and a common source of bugs stemming from unpredictable plug-in activation order.

1.5   What Can You Do With Eclipse?

The Eclipse platform is potentially useful for just about any software development task you can imagine. The Eclipse Foundation organizes its range of projects and sub-projects under what it calls the "Pillars of Eclipse." These include:

- *Enterprise development.* Tools and frameworks that span the entire software development lifecycle, including modeling, development, deployment tools, reporting, data manipulation, testing, and profiling. Projects under this pillar include: Business Intelligence and Reporting Tools (BIRT), Data Tools, Test and Performance Tools, and Web Tools.

- *Embedded and device development.* This is the area of immediate concern in this book. The projects under this pillar support building embedded applications as well as tools that assist with target management, device debugging, and building GUIs for mobile devices. These include: Device Software Development Platform, Embedded Rich Client Platform, Mobile Tools for Java, Native Application Builder, Target Management, and C/C++ IDE.

- *Rich client platform (RCP).* This is a platform for building and deploying so-called "rich client" applications with facilities for deploying native GUI applications to a variety of desktop operating systems, such as Windows, Linux, and Mac OSX. Under this pillar we find Equinox, a component framework based on the OSGi standard, along with the Plug-in Development Environment, Visual Editor, and the Eclipse platform itself.

- *Application frameworks.* A number of Eclipse projects provide frameworks that can be used as functional building blocks to accelerate the software development process. Unlike developer tools, application frameworks are deployed with the actual applications. Frameworks can be used either as standalone additions to Java applications, or can be leveraged as components on top of the Eclipse RCP. This supports the use of an integrated stack of open source frameworks on RCP to quickly build and deploy applications. Frameworks include: Eclipse Modeling, Graphical Modeling, Tool Services, Eclipse Communication, and Eclipse Process. BIRT and Data Tools from the Enterprise pillar are also included here.

- *Language IDE.* In addition to Java and C/C++, the Eclipse Foundation supports language IDE projects for Cobol and PHP. Third party plug-ins support a wide range of other languages such as MatLab, Ruby and Rails, Perl, and Python.

Summary

This chapter has been a brief introduction to what Eclipse is along with its history and current status. Eclipse is more than just an integrated development environment (IDE).

Instead, it is a framework for building IDEs. The Eclipse Platform provides a basic Graphical User Interface (GUI) on top of which plug-ins are added to provide functionality addressing a specific software development problem.

The next chapter will address the process of installing Eclipse on a workstation.

Resources

http://www.eclipse.org/—The official website of the Eclipse Foundation. There's a lot here and it's worth taking the time to look through it. Specific features of the website will be explained in more detail as we go along.

http://eclipse-plugins.2y.net/eclipse/index.jsp/—Eclipse Plugins. This site will give you a feel for the extent of the Eclipse ecosystem. It lists over 1000 plug-ins, both commercial and Open Source.

http://www.eclipseplugincentral.com/—Eclipse Plugin Resource Center and Marketplace. Not quite as extensive as Eclipse Plugins, this site lists some 400 plug-ins.

A Google search on "eclipse plugin" returns a great many hits, but except for the two sites listed above, all of the others seem to describe specific plug-ins mostly oriented toward Java and web development.

CHAPTER 2

Installation

2.1 System Requirements

The primary focus of this book is embedded software development using Linux. The primary focus of this chapter is installing and running Eclipse under Linux, which as we'll see, turns out to be fairly straightforward. Eclipse runs perfectly well under Windows and Mac OSX, and we'll take a look at the Windows installation process later in this chapter.

You will need a PC-class computer running a relatively recent Linux distribution. I happen to run both Red Hat Enterprise Linux (RHEL) 4 and Fedora Core 6, but feel free to use Debian, SUSE, Ubuntu, or whatever your favorite distribution happens to be. The Eclipse Foundation does caution, however, that Eclipse is only tested and validated on a "handful of popular combinations of operating system and Java Platform." From a Linux standpoint, the v3.4 Ganymede release has been validated on RHEL 4.0 and 5.0, and SUSE Linux Enterprise Server 10.

Installing a Linux distribution is beyond the scope of this book. There's lots of information and help available from the various distribution websites.

2.1.1 Hardware

Basic hardware requirements are relatively modest and largely dictated by Linux itself. For windowing operation, for example, Fedora Core 6 recommends a 400-MHz Pentium II or better, with 256 MB of RAM. The reality, of course, is that today anything less than a GHz processor and a GB of RAM is pretty much a doorstop anyway.

Storage requirements are likewise fairly minimal. The C Development Tools version of Eclipse that we'll be using takes approximately 70 MB of disk space.

2.1.2 Software

Since Eclipse is based on Java, you must have a Java Virtual Machine (JVM), also known as the Java Runtime Environment (JRE), available on your workstation. For RHEL 4.0, Eclipse recommends Sun Java 2 Standard Edition 5.0 Update 11 for Linux x86. Java 1.4.2 is also widely used and well tested in the Eclipse community.

Most contemporary Linux distributions install a JVM by default, but it may not be compatible with Eclipse. I found, for example, that the default JVM under RHEL 4.0 didn't work. Rather than take the time to puzzle out why, I simply downloaded another version that did work. We'll defer a discussion of downloading and installing the JVM until later, when we determine whether or not your default JVM works.

Finally, since our objective is to develop C programs for embedded devices, you'll need a GNU tool chain with the GCC compiler and linker and the GDB debugger. The tool chain is not always installed by default. Check to be sure it's there, and if not, follow your distribution's instructions for installing additional packages.

2.2 Obtaining Eclipse

Go to http://www.eclipse.org/ and click on the large orange button labeled **Download Eclipse**. This brings up a list of popular Eclipse packages consisting of the basic Eclipse platform plus one or more application add-ons. To the right of each list entry is a set of three links representing supported operating systems:

- Windows

- Linux

- Mac OS X

Click the **Linux** link for the "Eclipse IDE for C/C++ Developers." This brings up a list of mirror sites from which to download. Pick the one nearest you, understanding that in many cases it's not at all clear from a site's name where it is located geographically.

Download the tar.gz file to the directory in which you plan to install Eclipse. There are no hard and fast rules about where to install a given package. /opt is a good place. For reasons I can't explain now, I chose to install Eclipse in /usr/local.

2.3   Installation

Installation itself is trivial. Simply untar the `tar.gz` file that you downloaded. This results in the directory structure shown in Figure 2.1. It's not necessary to understand the content of these directories. Perhaps the most significant of them is `plugins/`, which contains all of the Java Archive (.jar) code. The `readme` directory contains a rather extensive HTML release notes file.

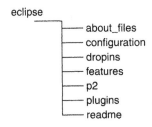

Figure 2.1: Eclipse directory structure.

The top-level `eclipse/` directory contains several files, the most important of which is the executable, `eclipse`. Eclipse is started by executing this file, either by double-clicking it in a graphical file manager window or by executing `/<path_to_eclipse>/eclipse` in a shell window.

If you're running a graphical desktop environment such as Gnome or KDE, you can create a custom launch button for Eclipse in the tool panel by following these instructions:

Gnome

1. Right-click on the tool panel.

2. Select **Add to Panel –> Custom Application Launcher**.

3. Fill in the pop-up dialog box:

Name: Eclipse 3.4

Generic Name: Eclipse

Comment:

Command: `/<path_to_eclipse>/eclipse`

4. Select an appropriate icon.

5. Click **OK**.

KDE

1. Right-click on the tool panel.

2. Select **Add** –> **Special Button** –> **Non-KDE Application**.

3. Fill in the pop-up dialog box:

Executable: `/<path_to_eclipse>/eclipse`

Optional command line arguments:

4. Select an appropriate icon.

5. Click **OK**

Then to start Eclipse, simply click on the launch button.

Go ahead and give it a try using any of the three mechanisms cited above. If you get to the screen shown in Figure 2.2, Eclipse is working. Click **Cancel** to terminate. You can skip the rest of this chapter and move on to the next, unless you want to try out

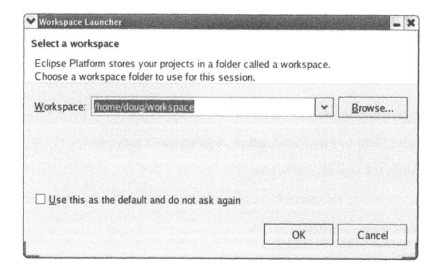

Figure 2.2: Workspace selection dialog.

Eclipse under Windows. If Eclipse failed to launch, it's probably because the JVM is either not present or not compatible. Continue with the next section.

2.3.1   Installing and Using a Java Virtual Machine (JVM)

JVMs can be downloaded from http://www.java.com/en/. Click on the **Free Java Download** button. This takes you to a Java Downloads for Windows page. Click on **All Java Downloads**. This brings up a page with downloads of the latest releases of Solaris, Linux, and Mac OS, as well as Windows. At the time this was being written, the latest release was Java 6 update 3. While Eclipse does not officially support this version, it does appear to work OK. Other versions of Java are available by clicking on **Other Java Versions**.

Download the Linux (self-extracting) file to the same directory where you downloaded Eclipse. The file is an executable, so execute it. You'll be asked to read and accept the Sun Microsystems Binary Code License Agreement for the Java SE runtime environment (JRE) version 6. The JVM is then extracted to `jre1.6.0_03/`. Note that if you choose to use a different version, the directory name changes accordingly.

To be sure the new version of the Java is the one that gets executed, it must appear in your path before the default version. You can add `/<path_to_jvm>/bin` to the beginning of your $PATH environment variable. Or you can create a link to the new Java executable in a directory that already appears in your `$PATH` ahead of the directory holding the default version.

Execute the shell command `whereis java` to determine where the default version is located. Then execute `echo $PATH` to find a suitable directory that appears earlier in your path. In my case, the default java is in `/usr/bin` and it turns out that `/usr/local/bin` shows up just ahead of that. So I put a link to `/usr/local/jre1.6.0_03/bin/java` in `/usr/local/bin`.

You can skip the rest of this chapter and move on to the next, unless you want to try out Eclipse under Windows.

2.4   Installing Eclipse Under Windows

Go back to the Eclipse downloads page, but this time select the **Windows** link and download the file to an appropriate destination directory. In this case the file is a `.zip` that must be opened with WinZip. Extract the `.zip` file to the directory of your choice. This results in almost the same directory structure as that shown in Figure 2.1. The `about_files/` subdirectory is missing.

Start Eclipse from a file manager window by double-clicking `eclipse.exe` in the `eclipse/` directory. If your Windows system has a JVM, and most likely it does, you should see a screen like Figure 2.3. For now, click **Cancel**.

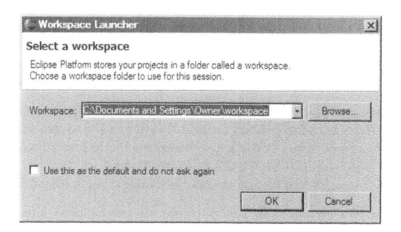

Figure 2.3: Workspace selection dialog under Windows.

You'll probably want to create a shortcut on the desktop for starting Eclipse. Right-click on `eclipse.exe` and select **Create Shortcut**. This creates a shortcut in the `eclipse/` directory. Drag that over to the desktop.

Note incidentally that Eclipse does not use the standard Windows program installation mechanism, and it doesn't put anything into the Windows registry. To uninstall Eclipse, simply delete the `eclipse/` directory.

2.4.1   Installing a JVM

If Eclipse did not start correctly, your system may not have a JVM. Go to http://java. com/en and click the **Free Java Download** button. Java offers two different mechanisms for installation under Windows—online and offline. Clicking the link **Windows XP/Vista/2000/2003 Online** downloads a small (360 KB) executable, `jre-6u3-windows-i586-p-iftw.exe`. This program in turn installs the rest of the JVM from the web.

You are given the opportunity to view, and then either to accept or to decline the terms of the Java license. Assuming you accept, the installation proceeds without any further user input required.

Alternatively, you can click **Windows XP/Vista/2000/2003 Offline**, which downloads a much larger (about 13.8 MB) executable, `jre-6u3-rc-windows-i586.exe`. This is the entire JVM package. The offline installation offers additional options—the Google toolbar and desktop—and more control over the installation process. A custom setup screen allows you to select the options to be installed. Unless you're an "advanced" user, it's probably best to accept the defaults.

Following installation, restart your browser and go back to http://java.com/en/. Select the **Advanced** tab and click the link **Do I have Java**? to bring up the Verify Installation page. Click the **Verify Installation** button. This should confirm that the JVM is properly installed. If not, you may need to configure the JVM.

Open the Windows Control Panel and double-click the **Java** icon (the coffee cup). Select the **Advanced** tab and click on the + next to **Default Java for browsers**. Check all the boxes on that branch to enable Java for the web browsers on your system.

2.5   Embedded Software Development on Windows

Remember that our objective here is to use Eclipse for developing software for embedded devices, with an emphasis on those that are Linux-based. While it is possible to do embedded development under Windows, it's somewhat more difficult because Windows, by itself, lacks a number of tools and services that are necessary, or at least highly desirable, for embedded software development.

Windows XP, other than the server edition, lacks network server facilities such as NFS (network file system) and TFTP (trivial file transfer protocol) that are very useful for debugging code on a target board. But most importantly, Windows lacks a tool chain for building software—a compiler, linker, assembler, libraries, etc.

The most widely used tool chain for embedded development is the GNU tool chain, which comes standard with just about every Linux distribution. There are two common approaches to adding a GNU tool chain to Windows: Cygwin and MinGW.

2.5.1 Cygwin

Cygwin is described as a "Linux-like environment" for Windows. It was originally developed in 1995 by Cygnus Solutions, which was subsequently purchased by Red Hat. Red Hat now maintains both the open source version and a licensable, proprietary version for people who want to maintain their own applications as proprietary.

Cygwin consists of two basic parts:

- A Windows DLL (cygwin1.dll) that acts as a Linux API emulation layer providing substantial Linux API functionality.

- A collection of tools that provide Linux look-and-feel. Among these tools is the GNU tool chain.

The primary motivation for Cygwin is to provide Unix/Linux functionality in a Windows environment, but it is not a way to run native Linux apps under Windows. Applications must be rebuilt from source to run in the Cygwin environment.

Nevertheless, it can be a useful tool for experimenting with C development with Eclipse under Windows. Note, however, that to build code for an embedded target, you will need a build of the GNU tool chain that supports your target processor. Many chip and board vendors provide Linux-based tool chains for their architectures, but rarely offer the tool chain built for Cygwin. So you will likely be on your own to build the target tool chain.

Another perceived drawback to Cygwin, for desktop applications anyway, is that the `cygwin1.dll` is released under the GPL. This means that anything that links with it, i.e., an application, is considered a "derivative work" and must itself be released under the GPL. On the other hand, this wouldn't be a problem for an embedded application intended for a target that runs real Linux. It is widely accepted that a Linux application running in user space and using only the published kernel APIs is not a derivative work.

Another nice feature of Cygwin is that it happens to include NFS and TFTP servers.

Installing Cygwin

Go to http://www.cygwin.com/ and click on the **Install Cygwin now** icon. There are several icons and links on this page that point to the same target, `setup.exe`.

When you click one of these links, Windows asks if you want to run or save the file. I generally save executables and then run them locally, but it's your call.

In either case, execute `setup.exe`. You may get a warning saying that the publisher could not be verified and asking if you really want to run it. Go ahead, it's safe. Following an initial information screen, you are offered three installation types:

- Install from Internet (default)

- Download Without Installing

- Install from Local Directory

The next screen lets you specify a root directory and select a couple of options. The recommended defaults for the options are good. Next you're asked to select a directory in which the downloaded packages will be stored. These are then available for subsequent reinstallation. Oddly, the default is the Desktop for the current user. I prefer to put stuff like this in the `\downloads` directory.

The next screen asks how you connect to the Internet. Select appropriately and continue. You are presented with a list of download mirror sites with the intention that you pick one geographically close to you. But again, most of the names offer no clue about where they may be located. Continuing to the next screen causes another setup program to be downloaded and you are presented with a package selection menu.

Sadly, this is not the most intuitive or user-friendly menu. Expand the **Devel** category by clicking on the +. The result is shown in Figure 2.4. Most of the packages are designated as "Skip," meaning they won't be installed. Scroll down to the gcc-core package and click on the word "**Skip**" in the New column. Skip changes to a version number and the **Bin?** column changes to a check box. This package, in its binary form, is now selected for installation. The **Src?** column is an open box giving you the option of downloading the source code as well. You might want to select this if you need to build a target version of gcc.

Scroll down and select gdb: The GNU Debugger as well. Again, you might want to check the Src? box if you will need to build a version for your target architecture. That should be all we need for C development. Now expand the net category and select the nfs-server and xinetd. Clicking **Next** starts the download. This is a lengthy process because we are, after all, building a fairly complete Linux environment. It's much more than the few packages we selected here.

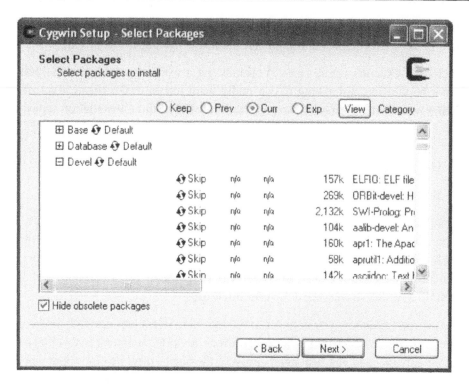

Figure 2.4: Cygwin package selection.

Following the download, you have the option of creating icons on the desktop or in the Start Menu. You're done. You'll find a folder named `cygwin\` in the folder specified for installation. It turns out to be the root directory of the Cygwin Linux environment. You'll also find a rather oddly named folder in the download directory that has a `setup.ini` file reflecting your package selection and a folder containing all of the download compressed package files.

Double-click the Cygwin icon and you'll get a bash shell as shown in Figure 2.5. Play around with some of the basic commands just to prove it really is a bash shell. Later we'll look at how to configure Eclipse to find the gcc compiler and other tools.

2.5.2   MinGW

MinGW, which stands for "Minimalist GNU for Windows," is the other popular approach to installing the GNU tool chain on Windows. The primary difference

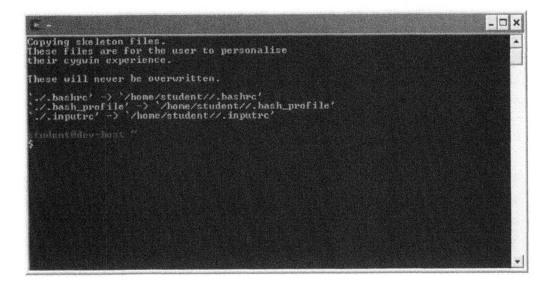

Figure 2.5: Cygwin bash shell.

between it and Cygwin is that MinGW uses the Windows C runtime libraries (mscvrt) instead of GNU's libc. This means that a compatibility layer is not needed, thus getting around the GPL issues associated with Cygwin.

Of course, this also means that MinGW generates native Windows code, which is fine for learning about and experimenting with CDT, but won't get you very far in building embedded target code. Nevertheless, the Eclipse documentation suggests that MinGW's direct support for the Windows environment provides the best integration with CDT.

MinGW is strictly an open source project and is hosted at http://sourceforge.net/index.php.

Installing MinGW

The MinGW download page is http://www.sourceforge.net/project/showfiles.php?group_id=2435. The first item in the list is MinGW-5.1.3. Clicking the **Download** button brings up another page with the actual file, MinGW-5.1.3.exe. Much like Cygwin, this is an installer that guides you through the installation. Start the program, select **Download, and install**. After agreeing to the license you can choose which package to install: **Previous, Current,** or **Candidate**. I recommend **Current**.

Select the MinGW base tools and the g++ compiler components (Figure 2.6). Select other compilers if you wish. Don't select MinGW Make. There's a more complete implementation of make called MSYS that you'll install in a subsequent step. Select an install location and a Start Menu folder, and click **Install**. There's about 60 MB to download, so it takes a while.

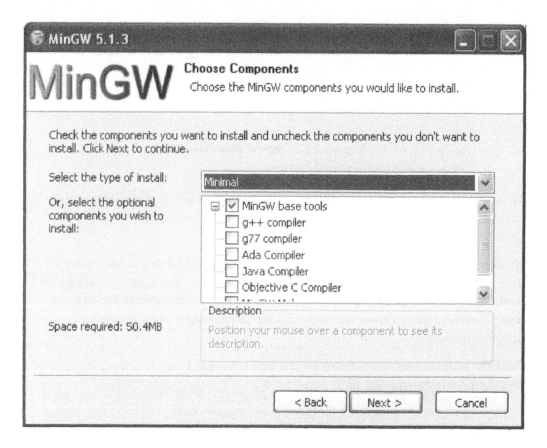

Figure 2.6: MinGW component selection.

Oddly, the MinGW installer doesn't install gdb. It can be downloaded at http://www.downloads.sourceforge.net/mingw/gdb-6.6.tar.bz2. Extract the contents of this file to the same location as MinGW.

If you want to create your own makefiles for use with CDT, you should install MSYS, also part of the MinGW project. MSYS, which stands for Minimal SYStem, is a POSIX-like command line interpreter (CLI) that serves as an alternative to the Windows command prompt, `cmd.exe`. As such, it facilitates the execution of POSIX-style build scripts and makefiles that are normally part of Open Source projects. The CLI is essentially a Bourne shell.

The MSYS installer is available from the same SourceForge page as MinGW. Click on MSYS Base System to get a list of the available releases. Select the Current Release and click on `MSYS-1.0.10.exe` to download the installer. After agreeing to the license terms and reviewing a release notes page, you get a dialog to select the installation folder. I chose to install MSYS in the `\MinGW` folder just to keep everything in one place. This is followed by a Select Components dialog, but in fact there is only one component.

Following installation, a command prompt window pops up to ask if you would like to execute the post-install process (Figure 2.7). Upon replying yes, "y", you're asked if you have MinGW installed and where it's located. Post-install then builds some script files.

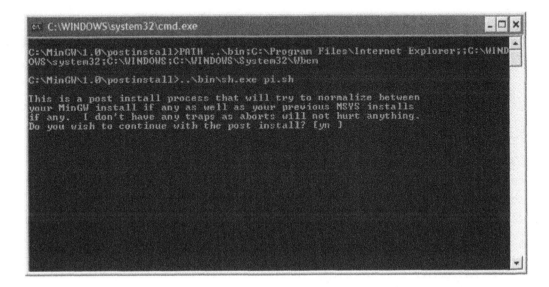

Figure 2.7: MSYS post install script.

The installation process puts an MSYS shortcut on your desktop. Double-click it to bring up the window in Figure 2.8. Try some POSIX shell commands to prove it works. For all practical purposes, MSYS provides the same functionality as Cygwin, so you really only need one or the other. In fact, its author describes MSYS as a fork of Cygwin that is "more friendly to the Win32 user."

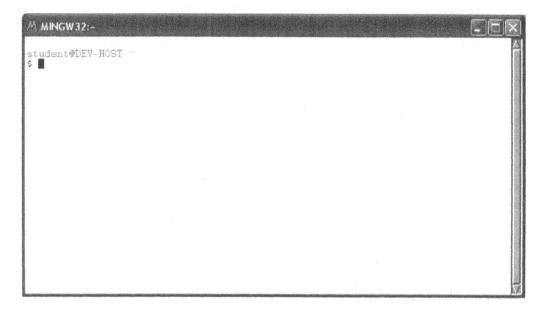

Figure 2.8: MSYS command shell.

2.5.3   NFS for Windows—nfsAxe

The combination of MinGW and MSYS does not include network server functionality such as NFS. There are several packages available, some for free, that add NFS server functionality to Windows. One that I've worked with is nfsAxe from LabF (http://www.labf.com/). It's a fairly extensive package that includes:

- NFS client
- NFS server

- FTP client

- Telnet

- LPD and LPR

- TFTP client

A free downloadable evaluation version of nfsAxe is available that supports one user and times-out after 30 minutes of operation. A commercial version supporting any number of users sells for $24 to $40 per user, depending on how many user licenses are purchased.

The download, `nfsaxe.exe`, is a self-extracting ZIP file. Use `nfsaxe.exe` to extract the package files and start an InstallShield wizard that steps you through the usual options of selecting an install directory, a setup type, and a program folder. Unlike Cygwin and MinGW, nfsAxe uses the standard Windows software installation process, so to remove it you must use the Control Panel Add/Remove software process.

Using nfsAxe creates a program folder with icons for all of its features, plus a user's manual and an uninstall process. The first time you double-click **NFS-Server** two things are likely to happen:

- A Windows security alert says the firewall has blocked the program from accepting connections. Click on **Unblock**.

- NFS_Server says the list of exported directories is empty, and asks, "Do you wish to create it?" Click **Yes** to bring up the window in Figure 2.9.

Click on **Add directory** to make one or more Windows directories visible to NFS. Then click on **Add User Access** to allow access to the directories you've exported. The simplest thing to do is select the wildcard, "*", for all the entries. Uncheck **Read only** if you want write access to the export.

2.5.4   Allegro—Another NFS Server

Allegro is a commercial product available from Franz, Inc., a web software tools vendor. A free 30-day evaluation of the $65 package is available from

Figure 2.9: NFS Server Settings.

http://www.nfsforwindows.com/home by sending them your name and email address. In return, they send you a link to a self-extracting ZIP file.

Following extraction, Allegro starts up a configuration utility (Figure 2.10) where you can specify exported directories and user access properties. An interesting feature of Allegro is that the exported name is separate and distinct from the path.

Allegro is a Windows service and by default it is started automatically.

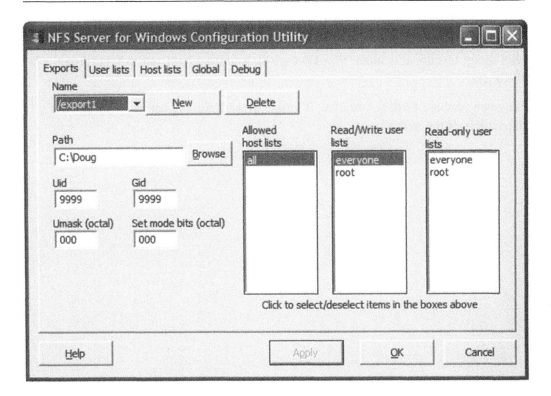

Figure 2.10: Allegro configuration utility.

Summary

This chapter covered the process of obtaining and installing Eclipse under both Linux and Windows. The Linux process is fairly straightforward. About the only hitch may be that the version of the Java Virtual Machine on your system may not be compatible with Eclipse.

Installation of Eclipse itself under Windows is equally straightforward. The problem, though, is that Windows lacks other features necessary to do software development, such as a compiler toolchain. Cygwin and MinGW are alternative approaches to installing the GNU toolchain under Windows.

Another useful tool for embedded development that Windows lacks is an NFS server. Cygwin includes an NFS server, but MinGW doesn't. Two packages that provide NFS server functionality under Windows are nfsAxe from LabF and Allegro from Franz, Inc.

Now that we have Eclipse installed, it's time to start playing around with it. That's the subject of the next chapter.

CHAPTER 3

Getting Started

3.1 Start Eclipse

Start in your home directory. There are three ways to start Eclipse, indeed to start any program under Linux running a graphical desktop environment:

- In a shell window, execute **<path_to_eclipse>eclipse**.

- In a file manager window, double-click the eclipse executable.

- Click on a custom launch button in the toolbar.

Eclipse always begins by asking you to select a workspace (Figure 3.1). The default workspace is the directory `workspace/` under your home directory. If the directory doesn't

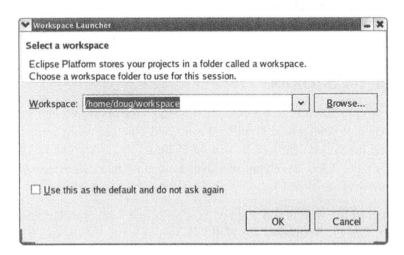

Figure 3.1: Workspace dialog.

exist, Eclipse will create it. If this is likely to be the only workspace you use, check the Use this as the default and do not ask again box to bypass this dialog.

If the workspace did not exist, Eclipse brings up a Welcome screen (Figure 3.2). This offers the opportunity to learn more about Eclipse before jumping right into it. Icons include:

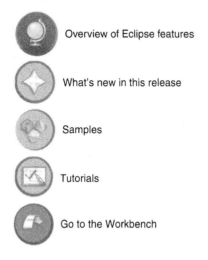

Overview of Eclipse features

What's new in this release

Samples

Tutorials

Go to the Workbench

The next time you start Eclipse in the same workspace the Welcome screen won't be displayed but you can always get back to it by clicking **Help** –> **Welcome**. For now click **Go to the workbench**, but feel free to come back to the samples and tutorials at any time.

3.2   Basic Concepts

At its core, Eclipse is really just a collection of tools for managing and manipulating files. The magic of course is in how these tools and other software components are structured and integrated. The user's view of Eclipse is a desktop known as a *Workbench*. Figure 3.3 is the empty workbench window that comes up before we've created any projects or files.

Across the top of the workbench window is a menu bar with familiar entries such as **File**, **Edit**, **Search**, **Window**, and **Help**, as well as some menus specific to Eclipse like **Refactor**, **Navigate**, **Run**, and **Project**. Below that is a tool bar whose icons may change depending on which perspective is visible and which view or

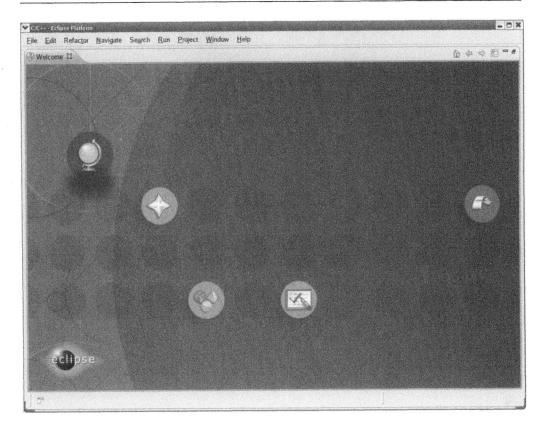

Figure 3.2: Welcome screen.

editor has the focus. The visible perspective is changed by clicking the ⬚ button at the far right of the tool bar. Eclipse CDT offers three default perspectives: **C/C++** that is currently visible, **Debug**, and **Team Synchronizing**. Two other perspectives are available by selecting the **Other**... button, **CVS Repository Exploring**, and **Resource**.

It is helpful to have a project open in order to discuss Eclipse basic concepts. Select **File –> New –> C Project**. In the **Project name:** field enter "hello". Under **Project types:** click the right arrow next to **Executable** and select **Hello World ANSI C Project** (see Figure 3.4). Click **Next** to bring up the Basic Settings dialog. Enter your name as the Author and change the Copyright notice and Hello world greeting, if you choose. Click **Finish**. The Project Explorer window on the left side of the workbench now shows some information about the hello project.

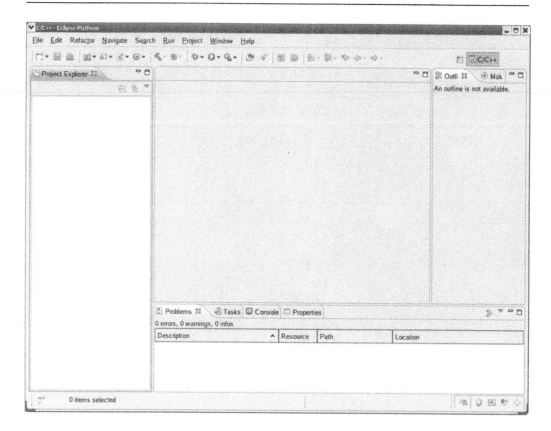

Figure 3.3: Empty workbench.

3.3   Perspectives, Editors, and Views

The Workbench window contains one or more Perspectives that are, in turn, collections of Views and Editors. A *Perspective* defines an initial set of views, and the layout of those views, to accomplish some specific task on a particular set of resources, or files. The workbench is currently displaying the C/C++ perspective typical of Eclipse CDT.

A workbench may have several perspectives open, but only one perspective is visible in a window. To make additional perspectives visible, open additional windows using the **Window –> New Window** command.

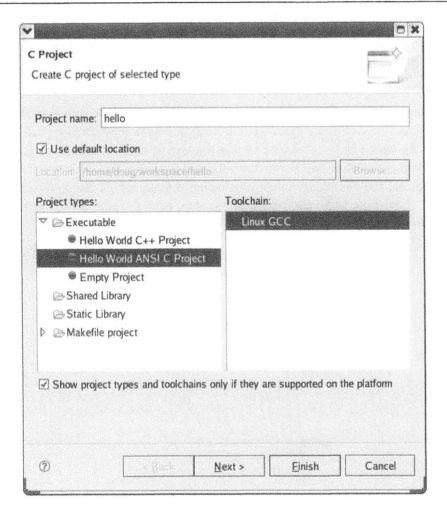

Figure 3.4: New project dialog.

The large space in the center of the workbench is the *Editor*. As you might expect, the editor allows you to open, modify, and save files. The editor window is the central feature of virtually all Eclipse perspectives. Different editors can be associated with different file types. Opening a file then starts up the corresponding editor, which may also change the contents of the menu and tool bars. The editor associated with C source files has a number of useful features that we'll look at shortly.

Multiple files can be open in the editor and are identified by tabs across the top of the editor window. An asterisk, "*", indicates the file has unsaved changes. Clicking the "X" icon to the right of the file name in the currently visible tab closes the file.

Views support editors and provide alternative presentations of the information in a project as well as ways to navigate that information. Views most often appear in tabbed stacks to the right and left of the editor window and sometimes beneath it. Icons on the right end of the tab bar allow the currently visible view in that stack to be minimized or maximized. Views also have their own menus represented by the down arrow icon at the far right of the view tab. Frequently used menu items may be represented by other icons in the tab.

Figure 3.5 is an example of the view menu for the Project Explorer view. The menu items are primarily concerned with how the view is displayed.

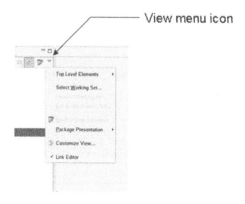

Figure 3.5: Project Explorer view menu.

A view can be moved around anywhere in the Workbench by dragging its title bar. As you move the view around, the mouse pointer changes to one of the *drop cursors* shown in Figure 3.6. This indicates where the view will be docked if you release the mouse. Try it with the Outline view on the right.

The default C/C++ perspective displays the **Project Explorer** view on the left and an **Outline** view on the right. The Project Explorer provides a hierarchical view of the

Drop cursor	Where the view will be moved to
⬆	Dock above: The view is docked above the view underneath the cursor.
⬇	Dock below: The view is docked below the view underneath the cursor.
➡	Dock to the right: The view is docked to the right of the view underneath the cursor.
⬅	Dock to the left: The view is docked to the left of the view underneath the cursor.
🗇	Stack: The view is docked as a Tab in the same pane as the view underneath the cursor.
⊞	Detached: The view is detached from the Workbench window and is shown in its own separate window
⊘	Restricted: You cannot dock the view in this area.

Figure 3.6: Drop cursor icons.

resources contained in a project. It allows for adding or importing new files or directories, deleting or exporting files, and opening files for editing.

The Outline view displays an outline of the structural elements of the file currently visible in the editor window. Since we don't currently have a file open in the editor, the Outline view is empty. Click the arrow to the left of the project name in the Project Explorer to expand the project. Then expand the src entry to reveal the file hello.c. Either double-click on hello.c or right-click and select **Open**. The source file shows up in the editor and there's now something in the Outline view on the right (Figure 3.7).

Note first of all that the editor, like any good programming language editor, employs syntax coloring. Click just to the right of the opening brace in main and note that the closing brace is highlighted. Try the same thing with the left parenthesis of the puts statement.

With the cursor located anywhere in the main function, the vertical bar on the left, known as the *marker bar*, shows the extent of main. The same thing happens if you click on main in the Outline view.

Roll your mouse over puts. A help window pops up showing the function declaration from the header file along with any comments associated with that declaration.

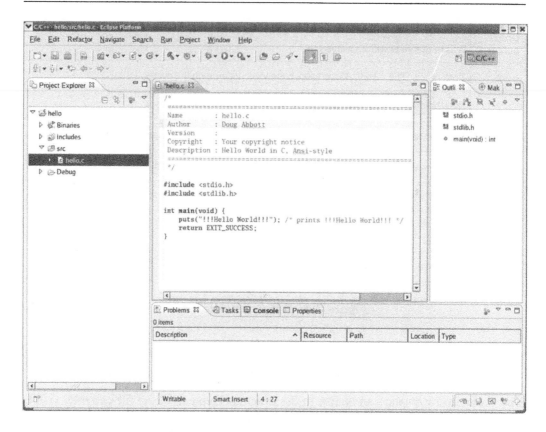

Figure 3.7: hello.c in editor window.

It's also worth noting that as soon as you created the hello project, Eclipse built it using the default gcc compiler. The results of the build are shown in the **Console** view below the editor. This brings us to a discussion of all four of the views that normally appear below the editor.

3.3.1   Problems View

If any errors or warnings are encountered in the course of building a project, they will be logged in the **Problems** view. Currently that view is empty because the project built successfully. It's easy enough to introduce an error, for example, by deleting the semicolon at the end of the puts statement.

Make that change and save the file. Note that by default, Eclipse does not automatically save any changed files before it builds a project. There is a preference option to save automatically before a build.[1] We'll look at preferences later in this chapter.

There are several ways to build the project. For now select **Project –> Build All**. The Problems view now shows a syntax error and tells us where it is. The error line is also identified in the editor with an icon in the marker bar. By default, problems are grouped by severity with different icons in the first column representing warnings and errors. If there are several items in the Problems view, clicking on an item moves the editor to the corresponding line, opening the file if necessary.

The Problems view can be filtered to show only warnings and/or errors for a particular resource or group of resources (Figure 3.8). Filters are accessed from the Problems view menu –> **Configure filters**... You can create multiple filters and enable and disable them as needed. Filters are "additive" so that any problem that satisfies at least one enabled filter will be shown.

Problems can also be sorted along several dimensions by selecting **Sort By** from the Problems view menu.

3.3.2 Tasks View

The **Tasks** view lets you create tasks related to the project and link those tasks to specific resources. There are several ways to add a task to the list. Right-click in the Tasks view and select **Add Task** to bring up the dialog in Figure 3.9. Here you can enter a description of the task and its priority. You can even check the task as completed although it's a little hard to understand why you would be adding a task that's already completed.

There are text boxes for entering an Element, Folder, and Location, but oddly enough you can't enter anything there. So tasks created by this method can't be linked to a resource. To create a task linked to a resource, `hello.c` for example, right-click in the marker bar on the left side of the editor window and select **Add Task**. Try it on the comment line that says "Copyright." The same dialog box comes up but now it's

---

[1] Personally, I think that should be the default.

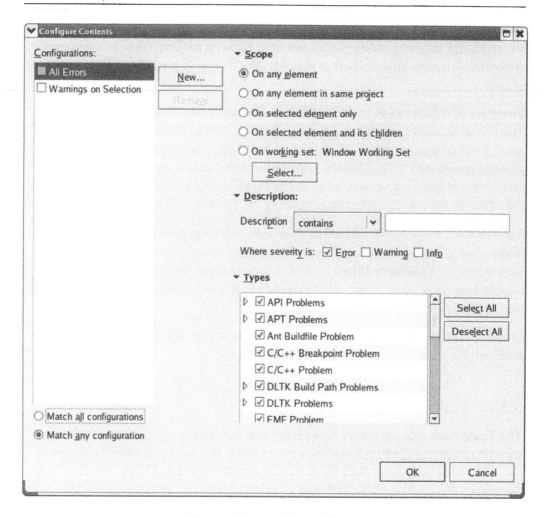

Figure 3.8: Problems filter.

labeled Properties instead of Add Task, and the Element, Folder, and Location fields are filled in.

Change the description to "update copyright" and click **OK**. A task icon appears in the marker bar at line 6. Like Problems, Tasks can be filtered to show only a relevant subset and can be sorted along several dimensions. Your Task view should look something like Figure 3.10.

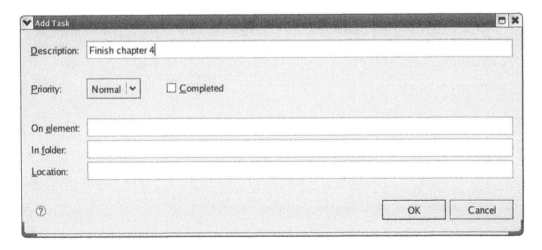

Figure 3.9: Add Task dialog.

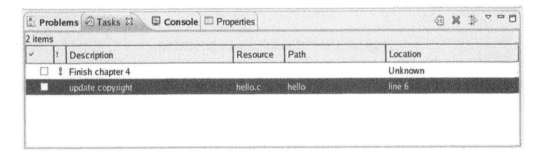

Figure 3.10: Task view.

3.3.3 Console View

The primary role of the **Console** view is to display program output, and output from the build tools. The Console view is connected to stdin, stdout, and stderr. Although several consoles may be open at any given time, only one is visible in the view. An icon on the tab bar lets you select the visible console.

Figure 3.11 shows two different console views (a and b). When a program is running in the console, a red rectangle icon appears that can terminate the program. Other icons remove terminated console views, clear the console, open a new console, and lock scrolling.

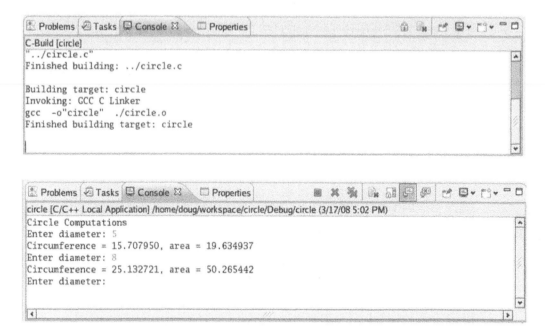

Figure 3.11: Console views (a and b).

The Console view only represents programs that are running on the host. Programs running on an external target will display their output in some other fashion, such as a terminal emulator window. In Chapter 6, "Device Software Development Platform," we'll look at an Eclipse project that makes remote programs visible in the Console view.

3.3.4   Properties View

Every object and/or resource in Eclipse has certain "properties," the natures of which depend on the type of object. The **Properties** view shows the properties of any object selected in one of the other views. With the Properties view visible, click on the project name "hello" in the Project Explorer view to bring up something like Figure 3.12.

Click on "hello.c" in the Project Explorer view to see a slightly different set of properties. The Properties view is read-only—you can't change anything here—and to be honest, it doesn't tell you a whole lot. A more extensive, editable view of properties is available from an object's *context menu* described later in this chapter.

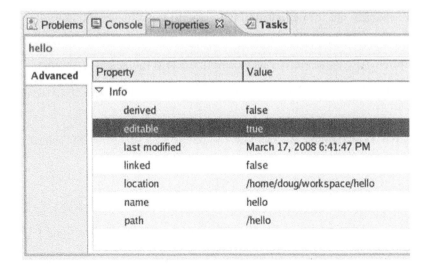

Figure 3.12: Properties view.

Thus far we've explored several of the more common views available in Eclipse. Later on we'll encounter other views more specific to the C development environment.

3.4 Menus

Like any good windowing program, Eclipse has a set of menus arrayed across the top of the main display window. Most of the items in these menus are fairly familiar, but a few deserve some additional description.

3.4.1 File Menu

This is a fairly standard file menu. Some additional items worth mentioning are:

- **Convert Line Delimiters To**: Changes how text lines are terminated for the selected file. Each of the three major operating systems that Eclipse supports has a different convention for how text lines are terminated:

  1. Unix (default): Line feed (0xa)

  2. Windows: Carriage return and line feed (0xd, 0xa)

  3. Mac OS/9: Carriage return (0xd)

The changes are immediate and persist until you change the delimiter again. It's not necessary to explicitly save the file.

- **Import**: Allows resources to be imported into the selected project.

- **Export**: Allows resources to be exported out of a project to some other location.

- **Switch Workspace**: Allows you to change to a different workspace. This restarts the workbench.

3.4.2   Edit Menu

The Edit menu also has many of the familiar options. Some Eclipse- and CDT-specific features include:

- **Incremental Find Next/Previous**: Search for expressions in the active editor. As you type the search expression, Eclipse incrementally jumps to the next/previous exact match.

- **Add Bookmark**: Adds a bookmark in the active file on the line where the cursor is displayed.

- **Add Task**: Adds a task in the active file on the line where the cursor is displayed.

- **Word Completion (Alt + /)**: Attempts to complete the word currently being entered in the active editor.

- **Quick fix**: Supposedly offers suggestions on correcting certain errors when the cursor is on a line that has an error. Unfortunately, the only thing I've seen so far is "No suggestions available."

- **Content Assist**: Opens a dialog at the current cursor location to offer assistance in the form of proposals and templates. The templates can be configured through the Window menu at **Window –> Preferences –> C/C++ –> C/C++ Editor –> Content Assist**.

- **Parameter Hints**: Displays the parameter portion of a function prototype.

- **Format**: Reformats a source file to match the currently selected coding style.

3.4.3 Refactor Menu

There's only one item in the Refactor menu: **Rename**. This is a way to rename a selected object and have the change propagated through the entire project.

3.4.4 Navigate Menu

As the name implies, this menu helps you navigate through a large project in a number of different ways. We'll look at many of these menu items in more detail in the next chapter, where we get into C programming:

- **Open Type Hierarchy**: Displays the Type Hierarchy view for the selected object, provided the object resolves to a defined type.

- **Open Call Hierarchy**: Displays the Call Hierarchy view for the selected function. The Call Hierarchy can show which functions call this function, and which functions this function calls.

- **Open Declaration**: Opens the declaration of the selected object: a function, variable, class, type, etc.

- **Open Resource**: Displays a dialog allowing you to select any resource in the workspace to open in an editor.

- **Last Edit Location**: Moves the cursor to the line that contains the most recent edit.

- **Go to Line**: Displays a dialog to specify a line number to which to move the cursor.

3.4.5 Search Menu

The Search menu offers three different ways to search for text, represented by four menu items. Figure 3.13 shows the C/C++ search dialog that allows you to search for text in specific language elements. The scope of the search can be the entire workspace or the set of resources selected in the Project Explorer view. The results of the search appear in the Search view.

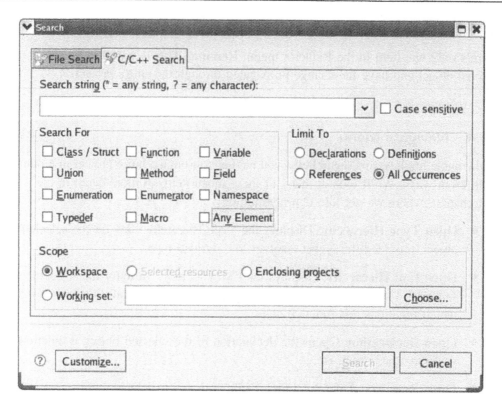

Figure 3.13: C/C++ Search dialog.

Figure 3.14 shows the File search dialog. Here you can search for a text string, a regular expression, and/or a file name pattern. Again, the search can encompass the entire workspace or only selected resources, and the results of the search appear in the Search view.

Finally, with the cursor set in any word, you can select **Search –> Text** and immediately see the Search view with a list of the files containing the selected text.

3.4.6   Project Menu

The Project menu is concerned, perhaps not surprisingly, with managing projects, which is in fact the primary organizing principle of Eclipse. The menu includes items for opening and closing projects, building projects, and making specific targets.

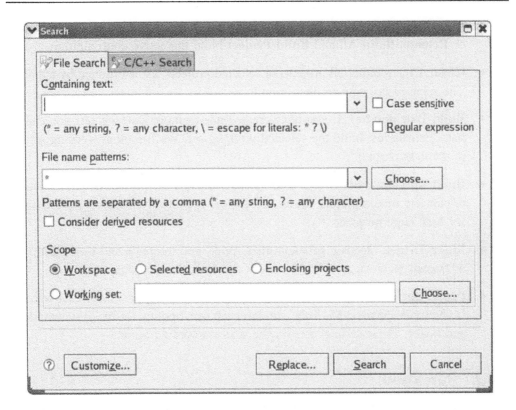

Figure 3.14: File Search dialog.

Select the hello project in the Project Explorer view and click the Project menu. Menu items include:

- **Close Project**: Closes the selected project(s). Any project files open in editors are closed and the hierarchy in the Project Explorer view is collapsed.

- **Open Project**: A project that was previously closed can be opened. Note that the Project Explorer view must have the focus for Open and Close to be active.

- **Build All**: Builds all projects in the workspace. Currently there's only one, the "hello" project. Clicking **Build All** now will probably result in the message "Nothing to be done for 'all'" since the project has already been built. This is a full build, that is, all files are built.

- **Build Project**: Builds the project currently selected in the Project Explorer view. This is a full build.

- **Build Configurations**: Allows you to set the active configuration, either **Debug** or **Release. Build All** and **Build Project** build the active configuration.

- **Clean**: Cleans either all projects or selected projects, as determined by a dialog. The selected projects are then rebuilt.

- **Build Working Set**: If you have created one or more *working sets* you can have Eclipse just build the selected working set. We'll look at working sets in the next chapter.

- **Build Automatically**: If this item is checked then the project is automatically rebuilt any time a project file is saved. This is probably not a very good idea for very large projects.

- **Make Target**: Opens a submenu that allows you to create and then build additional make targets. We'll look at this is in the next chapter.

- **Properties**: Opens a rather extensive properties dialog, an example of which is shown in Figure 3.15. This particular tab sets options for the gcc compiler and linker. These settings are for the selected configuration.

3.4.7   Run Menu

Having built a project, the next thing we probably want to do is run or debug it. Eclipse runs projects from a *launch configuration* that specifies the program to run, its arguments and environment, and how the program connects to Eclipse (i.e., is it a local process or is it running remotely on a target board of some sort?). A launch configuration also specifies the debugger and how it connects to Eclipse.

Before we run or debug a project, we must first create a launch configuration. Click on **Run –> Run Configurations**... Click the **New** button, the one farthest to the left in the left-hand panel, to create a new configuration under "C/C++ Local Application." This brings up the dialog shown in Figure 3.16.

The configuration name has been automatically set to match the project name. You need to select an application to run by clicking **Search Project**... and then selecting `/hello/Release/hello`. Click the other tabs just to see what's there. The hello program takes no arguments and has no environment variables. We'll look at the **Debugger** tab in the next chapter.

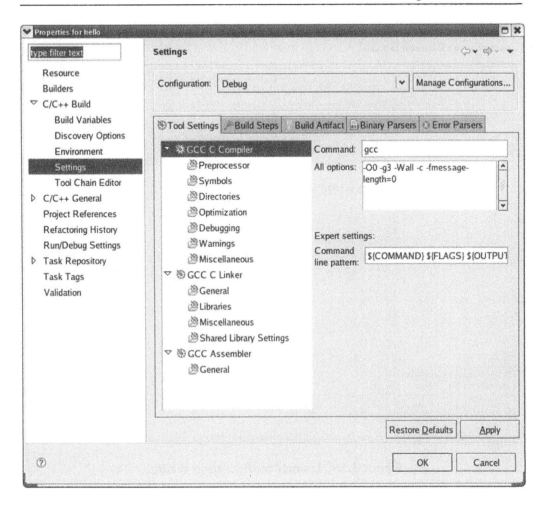

Figure 3.15: Project properties dialog.

Now click **Apply** and then click **Run**. The program output appears in the Console view. With at least one launch configuration established, the other Run menu actions become meaningful:

- **Run**: Rerun the most recent launch in Run mode.

- **Run History**: Presents a submenu of configurations launched in Run mode. Currently there's only one.

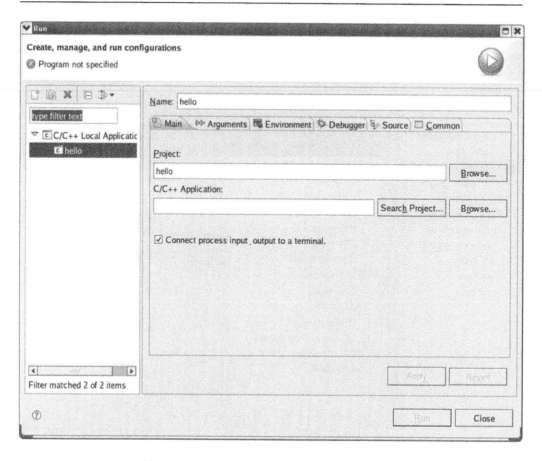

Figure 3.16: Launch configuration dialog.

- **Run As**: Presents a submenu with one item: "Local C/C++ Application."
  Clicking that brings up a dialog where you can select either the release or debug
  binary to run.

- **Debug**: Rerun the most recent launch in Debug mode.

- **Debug History**: Presents a submenu of configurations launched in Run mode.
  Currently there's only one.

- **Debug As**: Presents a submenu with one item: "Local C/C++ Application." Clicking
  that brings up a dialog where you can select either the release or debug binary to run.

- **Debug Configurations**...: Opens the launch configuration dialog for debug mode configurations. The hello configuration we created for run mode is selected and can be used as-is. We'll explore debugging in the next chapter.

The Run menu also includes a number of execution control actions that are used for debugging.

3.4.8   Window Menu

This menu offers a number of options for selecting views and perspectives and moving around the workbench. Actions include:

- **New Window**: Opens a new workbench instance. This allows you to have two or more perspectives visible at once.

- **New Editor**: Opens an empty editor window.

- **Open Perspective**: Allows you to select another perspective. The choices offered depend on where you are. In any case, there's always an **Other**... case that lists all of the perspectives available.

- **Show View**: Brings up a list of views to make visible. Depending on where you are, the views you are most likely to be interested in are listed first. Again, there's an **Other**... selection to select any possible view.

- **Customize Perspective**: Each perspective includes a predefined set of actions accessible from menus and the workbench toolbar. This command brings up a dialog that lets you customize the actions in the current perspective.

- **Save Perspective As**: Saves the current perspective, thus giving you the opportunity to create custom perspectives, which may be opened with **Window –> Open Perspective –> Other**.

- **Reset Perspective**: Returns the current perspective to its original layout.

- **Close Perspective**: Closes the current perspective. Basically, this just makes another open perspective visible.

- **Close All Perspectives**: Closes all open perspectives. This leaves the workbench essentially blank with just the **Open Perspective** icon visible. All open editors are closed.

- **Navigation**: Brings up an extensive submenu that provides another way to move among views, perspectives, and editors.

- **Preferences**...: Leads you to the configuration dialog for Eclipse. There are many preference options, which are the subject of the next section.

3.4.9   Help Menu

Eclipse comes with extensive help documentation that can be accessed in different ways:

- **Welcome**: Displays the welcome screen with access to tutorials and examples.

- **Help Contents**: Opens a new help window. A navigation panel on the left lets you browse through two user guides: the *Workbench User Guide* and the *C/C++ Development User Guide*. There's also a search window.

- **Search**: This is one of several help actions that pops up a Help view in the current perspective. See Figure 3.17.

- **Dynamic Help**: Brings up the Help view as another way to browse the help files.

- **Key Assist**...: Displays a pop-up window with all of the shortcut keys.

- **Tips and Tricks**...: Displays some helpful ideas for improving productivity in the separate help window.

- **Report Bug or Enhancement**...: Provides a convenient mechanism for filing bug reports.

- **Cheat Sheets**...: These are short tutorials that display in the Cheat Sheets view.

- **Software Updates**: Finds and installs updates to Eclipse software. You have a choice between updating only currently installed features, and searching for new features to install.

- **About Eclipse Platform**: The usual "about" type of information. Buttons provide additional information about features, plug-ins, and configuration details.

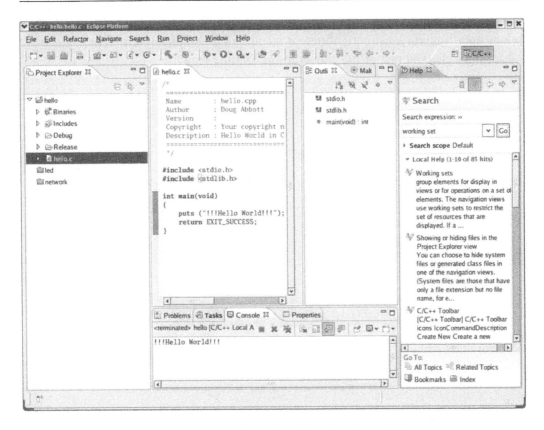

Figure 3.17: Workbench with Help view visible.

3.4.10   Context Menu

Again, like any good windowing program, Eclipse offers several different ways to invoke its functionality. In addition to the menus described above, many Eclipse actions are available from the toolbar just below the menu bar. The actions available in the toolbar may change depending on which perspective is visible and which view has the focus.

In addition, just about every object has a *Context menu*, accessed by right-clicking on the object. The Context menu includes a collection of actions derived from the other menus that are commonly performed on the selected object.

For example, right-click on the project name, "hello," in the Project Explorer to bring up the menu shown in Figure 3.18. This includes actions from the File, Project, and Run menus that are useful to perform at the project level. Right-clicking on hello.c brings up a similar, yet slightly different menu. For a completely different Context menu, right-click on one of the entries in the Outline view.

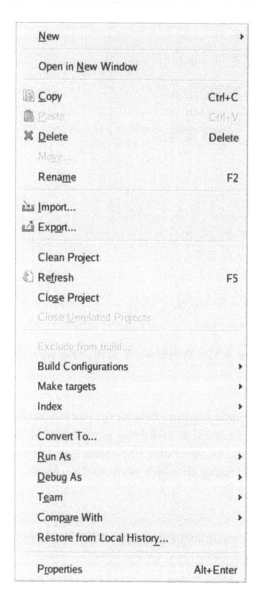

Figure 3.18: Project-level Context menu.

Browse around different views and different objects to get a feel for the range of the Context menus.

3.5 Configuring Eclipse

Perhaps not surprisingly, Eclipse offers an extensive array of configuration options that let you tailor its appearance and operation to your particular tastes or to company development standards. The Preferences dialog shown in Figure 3.19 is activated with **Window –> Preferences**.

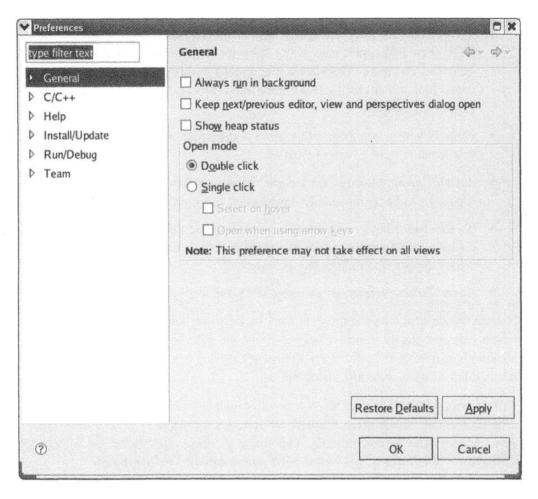

Figure 3.19: Preferences dialog.

As shown here, preferences are arranged into six major categories. At this point, it's not worth going through an exhaustive explanation of all the preference options. Many of them are somewhat obscure, in any case. There is an extensive section in both *the Workbench User Guide* and the *C/C++ Development User Guide* that describes all of the preference options in detail.

Briefly, the six major categories are:

- *General.* Pretty much what it says. There are options here for managing the appearance of Eclipse, configuring editors and shortcut keys, startup and shutdown operations, and other features.

- *C/C++.* This category manages a large number of configuration options for CDT, including appearance, editor behavior, environment variables, debug behavior, and more. There are a few options in this category that are worth a closer look in this chapter (see below) and others that we'll take a closer look at in the next chapter.

- *Help.* Specifies how help information is displayed and offers a way to include help content from a remote location.

- *Install/Update.* Manages the Eclipse software update and installation process. Updates can be automatically downloaded and installed.

- *Run/Debug.* Manages the process of running and debugging code under Eclipse, including things like console appearance, breakpoint behavior, and what happens when a program is launched.

- *Team.* Offers preferences for using CVS and team synchronization.

Note the circle icon with the question mark in the lower left corner. Clicking this icon brings up a context-sensitive help screen to the right of the dialog. This is, in fact, a common feature of virtually every Eclipse dialog and is a useful way to get more detailed information about the dialog options.

At this point it would be useful to browse around the preferences menus to get a better feel for what's there. Earlier, we saw that by default, Eclipse does not save changed files before building. The option to change that behavior is in **General –> Workspace**. It's called **Save automatically before build**.

3.5.1 C/C++ Preferences

Code Style

The CDT text editor incorporates "smart typing" features, that include things like auto-indentation and formatting, that are controlled by the Code Style preference shown in Figure 3.20. You can select from among four built-in code styles that include:

- K&R

- BSD/Allman

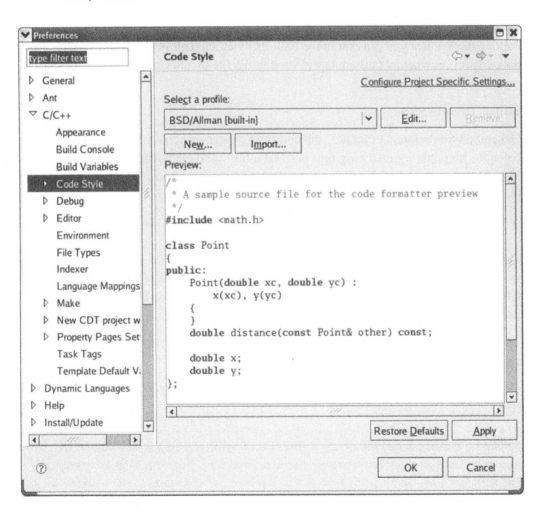

Figure 3.20: Code Style preference.

- GNU

- Whitesmiths

The primary difference among the four seems to be the location and indentation of opening and closing braces. The default is K&R, where the opening brace is on the same line as the expression or key word that introduces the block. This seems to be the preferred style among Linux programmers.

For what it's worth, my personal preference is BSD/Allman, where the opening brace is on the next line and indented to line up with the introductory expression. There are options to edit the built-in styles, import a style, and to create a completely new one.

Among the features you can edit in the built-in styles are **Tab policy** and **Tab size**. Tab policy specifies whether tabs are represented in the file by tab characters, 0x9, or spaces.

Editor Preferences

General editor preferences include the ability to change various colors. I find the default color for highlighting matching braces to be a bit "subtle." To change it, select **Matching brackets highlight** under **Appearance color options**: and then click on the **Color**: button.

A few of the more interesting editor preference subcategories are described here.

Content Assist   With Content Assist, the C/C++ editor can offer suggestions about key words and phrases commonly used in C. Type part of a keyword followed by **Ctrl+Space** to bring up a list of suggestions. When the list has been reduced to one item, the editor can automatically insert that item into your code.

The related auto-completion feature can use ".", "->", and "::" as triggers to invoke auto-completion on structures and class definitions.

Content Assist is based on *templates* that can be specified under Template preferences (see below).

Folding   *Folding* hides the detailed contents of selected regions in a source file. This can be a useful strategy for browsing through large files. Figure 3.21 is an example of a file with functions and macros folded. Click on the "+" icon next to the name to see the contents of a specific function or macro.

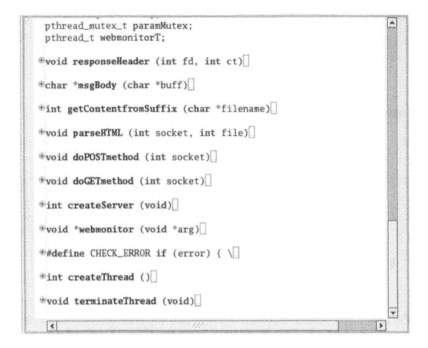

Figure 3.21: Folded source file.

Folding preferences let you select if folding is enabled when a new editor is opened, and which sections of code will be folded.

Syntax Coloring   There are a large number of semantic elements to which specific coloring and font styles can be assigned. Many of these are not enabled by default and among those that are enabled many share the same coloring. Here's your chance to "go wild," if you're so inclined, and create some exciting color schemes.

Templates   *Templates* are the basis for Content Assist. There's a default set of templates representing common C/C++ code snippets. You can add your own templates with **New,** and **Edit** existing templates.

Typing   Typing preferences are another way in which the editor provides assistance. You can tell the editor to automatically close quoted strings, parentheses, brackets, and braces. When you type an opening parenthesis or bracket, the editor automatically inserts the matching close and positions the cursor between them. Type what should be enclosed, then hit **Enter** to position the cursor just beyond the matching close.

Summary

This chapter has been a quick tour of basic Eclipse concepts and operation with an emphasis on the C/C++ Development Tools (CDT) environment. We began by examining the relationship among Perspectives, Views, and Editors. A Perspective is a collection of Views and Editors organized to accomplish a specific function. Editors allow files to be opened, modified, and saved. Views support Editors by providing additional information and functionality. The selection and arrangement of Views in a Perspective can be changed at will.

The basic functionality of Eclipse, like all windowing programs, is embodied in a set of menus. We looked in detail at many menu, or action, items that are specific to Eclipse. Many of these action items also appear in the tool bar and in Context menus. Items in the tool bar may change or become active or inactive depending on which view has the focus.

Finally, we looked at some of the many configuration and customization options available in Eclipse with particular attention to some of the interesting features of the CDT editor.

The next chapter goes into more detail about the CDT environment and the nature of projects.

CHAPTER 4

C/C++ Developers' Toolkit (CDT)

With a good basic understanding of Eclipse concepts as a background, this chapter looks at the C Developers' Toolkit in more detail. It covers how to create, configure, and build projects.

4.1 Obtaining the Sample Source Code

Source code and data files for all the examples in this book are available from http://www.intellimetrix.us/downloads.html in the file `EclipseSamples.tar.gz`. Download this file to your home directory and untar it. You'll find a new directory called `EclipseSamples/` with subdirectories for each of the projects described in the book.

4.2 Creating a New Project

For this exercise we'll create a fairly straightforward record sorting application. The records to be sorted consist of a name and an ID number. To simplify the code a bit, we'll replace any spaces in the name field with underscores. The program will sort a file of records in ascending order, either by name or ID number, as specified on the command line thusly:

```
record_sort <datafile> [1 | 2]
```

Where "1" means sort by name and "2" means sort by ID. Sort by name is the default if no sorting argument is given.

In the Eclipse C/C++ perspective, create a new **C Project** and call it "record_sort." The project type is **Executable** and we'll use the default workspace location. Clicking

Next brings up the Select Configurations dialog where you can select either or both of the Debug and Release configurations. Later you'll have the choice of building either of these configurations. The primary difference between them is that the Debug configuration is built with the compiler's debug flag, "-g", turned on to provide information to GDB. The Release configuration leaves out the debug flag.

There's also an **Advanced settings**… button to delve deeper into project configuration. We'll look at that later.

When you click **Finish** in the New Project Wizard, the workbench looks like Figure 4.1. The only item under the record_sort project in the Project Explorer view is Includes, which is a list of paths to search for header files. Eclipse attempted to build the project, but of course there's nothing to build since there are no source files.

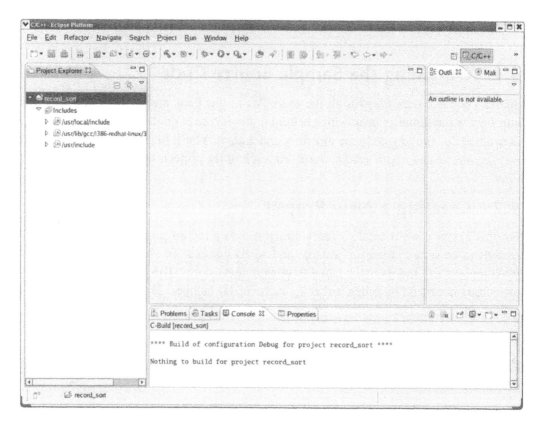

Figure 4.1: Empty project.

At this point it would be useful to take a look at the directory `workspace/record_sort/`. It contains just two files, `.cproject` and `.project`, both of which are XML code describing the project. The `.project` file provides configuration information to the base Eclipse platform while the more extensive `.cproject` file provides information to CDT. It's not necessary to understand the contents of these files, but it is useful to know they exist.

4.3   Adding Source Code to the Project

There are basically two ways to add source code to a C project. You can, of course, create a new file in an Editor window, or you can *import* existing files into the project. Execute **File > Import**... to bring up the Import Select dialog. Expand the General category and select File System. Click **Next**, then click **Browse**, and navigate to `Home/EclipseSamples/record_sort` and click **OK**.

This brings up the Import dialog shown in Figure 4.2. Select all three files and click **Finish**. Those files now show up in the Project Explorer view. Note that there is no `record_sort.c file`. That's because you're going to type it in yourself to get some practice with the CDT editor.

Click the down arrow next to the **New** icon at the left end of the tool bar and select **Source File** from the drop down menu. Name it "record_sort.c." An Editor window opens up with a preliminary header comment. The contents of `record_sort.c` are given in Figure 4.3, but don't start typing until you read the next section.

4.3.1   Content Assist

The CDT Editor has a number of features to make your coding life easier. These fall under the general heading of *Content Assist*. The basic idea of Content Assist is to reduce the number of keystrokes you must type by predicting what you're likely to type based on the current context, scope, and prefix. Content Assist is invoked by typing **Ctrl+Space** and it's also auto-activated when you type ".", "->", or "::" following a `struct` or `class` name.

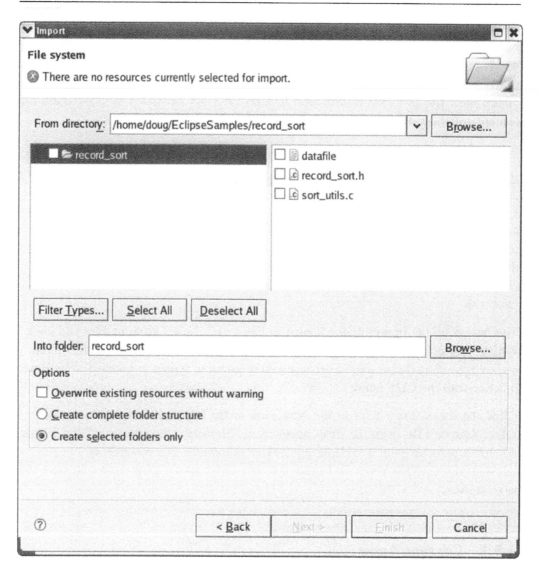

Figure 4.2: Import dialog.

4.3.2   Code Templates

Code Templates are an element of Content Assist that provide starting points for frequently used sections of code. When you enter a letter combination followed by **Ctrl+Space**, a list of code templates that begin with that letter combination is displayed. Select one and it is automatically inserted into the file at the current cursor location.

```
/*
 * author Doug Abbott
 *
 * Simple demonstration of building and running a project under
 * Eclipse.
 *
 * Usage:
 *     record_sort <filename> [1 | 2]
 *
 * Sorts records in <filename> either by name (1) or ID (2).
 * Default is name.  Outputs sorted file to stdout.
 */
#include <stdio.h>
#include <stdlib.h>

#include "record_sort.h"

int main (int argc, char **argv)
{
      int size, sort = 1;
      record_t *records;

      if (read_file (argv[1], &size, &records))
      {
            printf ("Couldn't open file %s\n", argv[1]);
            exit (1);
      }
      if (argc > 2)
            sort = atoi (argv[2]);

      switch (sort)
      {
            case 1: sort_name (size, records);
                  break;

            case 2: sort_ID (size, records);
                  break;

            default:
                  printf ("Invalid sort argument\n");
                  return_records (size, records);
                  exit (2);
      }
      write_sorted (size, records);
      return_records (size, records);
      return 0;
}
```

Figure 4.3: record_sort.c.

Try this: after entering the #include lines in Figure 4.3, type "ma" **Ctrl+Space**. This brings up a template for the main() function. Note that the format of templates is independent of whichever Code Style you've selected and the default is K&R. You can edit templates at **Window –> Preferences –> C/C++ –> Editor –> Templates**. We'll take a closer look at that later in the chapter. Other aspects of Content Assist can also be customized under Preferences.

4.3.3 Automatic Closing

As you type, note that whenever you type an opening quote ("), parenthesis [( ], square ([ ) or angle (<) bracket, or brace ({), the Editor automatically adds the corresponding closing character and positions the cursor between the two. Type whatever is required in the enclosure and hit **Enter**. This positions the cursor just beyond the closing character. However, if you move the cursor out of the enclosed space, to copy and paste some text for example, the **Enter** key reverts to its normal behavior of starting a new line.

In the case of an opening brace, the closing brace is positioned according to the currently selected coding style, and the cursor is properly indented.

Finally, notice that as you type, appropriate entries appear in the Outline view identifying header files, functions, and if we had any, global variables.

4.4 The Program

Before moving on to building and running the project, let's take a closer look at what it actually does. main() itself is pretty straightforward. It's just a set of calls to functions declared in sort_utils.c that do the real work.

The function read_file() reads in a data file that is assumed to be organized as one record per line, where a record is a text name and a numeric ID. It allocates memory for an array of records and a separate allocation for each name field.

There are two sort functions: one to sort on the name field, and the other to sort on the ID field. Both of these implement the shell sort algorithm, named after its inventor Donald Shell. Shell sort improves performance over the simpler insertion sort by comparing elements separated by a gap of several positions.

After the record array is sorted, `write_sorted()` writes it to `stdout`. This could be redirected to a file, of course.

The final step is to return all of the allocated memory in the function `return_records()`.

The program does virtually no "sanity checking" and, if you're so inclined, you might want to build some in. There's also very little in the way of error checking.

4.5 Building the Project

Once you've completed and saved the `record_sort.c` file, the next step is to build the project. All files that are created in, or imported into, a project automatically become a part of it and are built and linked into the final executable.

In the Project Explorer view, select the top-level `record_sort` entry. Then execute **Project –> Build Project** or right-click and select **Build Configurations –> Build All**. In the former case, the *Active Configuration* will be built. By default this is the Debug configuration. The Active Configuration can be changed by executing **Project –> Build Configurations –> Set Active**.

In the latter case, both the Debug and Release configurations will be built. In either case one or two new entries will show up under `record_sort` in the Project Explorer view. These entries represent subdirectories called `Debug/` and `Release/` that hold, respectively, the object and executable files for the corresponding build configurations. Each also contains a makefile and some Eclipse-specific files.

Initially the build will fail because some compile-time errors and warnings have been built into `sort_utils.c`. Open the Problems view, expand the Errors entry, find the item that says "structure has no member named 'Id'" and right-click on it. Select **Go To** to open `sort_utils.c`, if it isn't already open, and highlight the line that has the error (Figure 4.4). The correction should be fairly obvious.

Eclipse CDT identifies build problems several different ways. In the Project Explorer view, any project and source files that have problems are flagged with a red "X" icon for errors or a yellow shield icon with a "!" to indicate a warning. When a source file with errors or warnings is opened in the Editor, the tab shows the same symbol. The Outline view then uses the same icons to show which functions in the file have either errors or warnings.

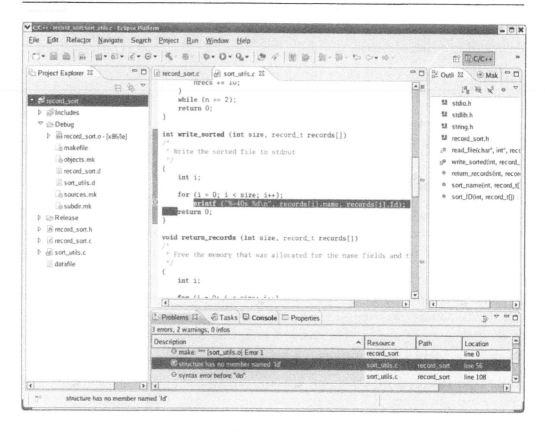

Figure 4.4: Identifying build problems.

The Editor window uses the same icons to identify the line on which each error or warning occurs. You can scroll through a source file and click on a line displaying either a warning or error icon and the Problems view will jump to the corresponding entry. If you roll the cursor over a line that's identified as an error or warning, the corresponding error message pops up.

Correct the problems and build the project again. This time the build should succeed and you'll see an executable file show up in the Debug tree in the Package Explorer view.

4.6   Debugging the Project

Execute **Run –> Debug**. If you executed **Build Configurations –> Build –> All**, you'll be asked to choose a local application to debug. Select the binary with the "bug" icon and click **OK**. Next you're asked if you want to open the Debug perspective, shown in Figure 4.5. Yes, you do. This may also automatically create a debug launch configuration as discussed in the previous chapter. If not, the launch configuration dialog will pop up and you can create a new configuration named record_sort under **C/C++ Local Application**.

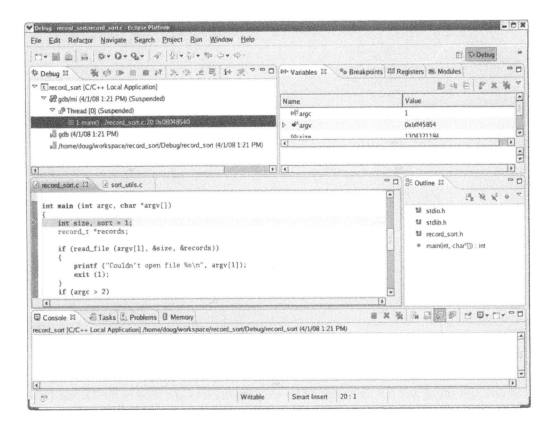

Figure 4.5: Debug perspective.

If the launch configuration dialog did not come up, execute **Run –> Debug Configurations**... because we need to modify the configuration in any case. Select the Arguments tab as shown in Figure 4.6 and enter "datafile" into the Program arguments: window. `datafile` is a set of sample data records for the program to sort. It was imported into the project along with `sort_utils.c` and `record_sort.h`.

Note that the first line of `main()` is highlighted in the Editor window and a green arrow in the marker bar identifies this as the current execution point. That's because, by default, Stop on startup at: main is selected in the launch configuration.

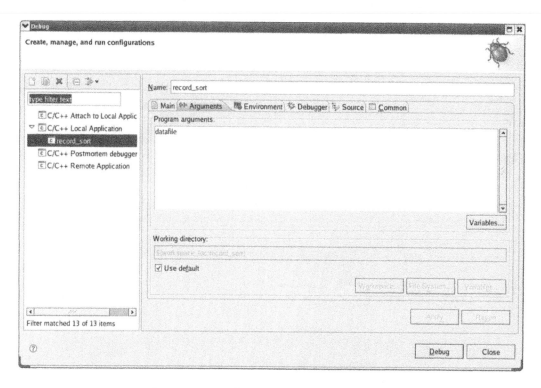

Figure 4.6: Debug launch configuration.

For now, go ahead and run the program by clicking the Resume icon in the Debug view toolbar. Hmmm, we didn't get exactly the results we expected. `datafile` has twelve records, but only one record is output to the Console view. That's because a couple of run-time errors have been built into the program to offer some practice using the debugger.

4.6.1   The Debug View

In the Debug view, right-click on the top level project entry and select **Relaunch** to start another debug run. The Debug view, shown in Figure 4.7, displays the state of the program in a hierarchical form. At the top of the hierarchy is a *launch instance*, that is, an instance of a launch configuration identified by its name. Below that is the debugger instance, identified by the name of the debugger, in this case, gdb. Beneath the debugger are the program threads under the debugger's control. For `record_sort` there is just one thread. In the next chapter we'll see how gdb/Eclipse handles multi-threaded programs.

Finally, at the lowest level are the thread's stack frames, identified by function name, source code line, and program counter. Currently there is only one stack frame for `main()`, stopped at `record_sort.c`, line 20.

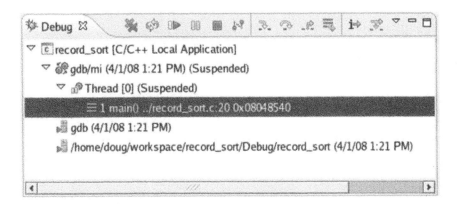

Figure 4.7: Debug view.

The Debug view's tool bar has lots of buttons for managing program execution as shown in Table 4.1.

The Debug view tool bar also has a menu with two items:

- **Show Full Paths**: Toggles between showing the full path to the source files and just the file name. In practice, most of the time the path is just ../ anyway.

- **View Management**: The Debug view can automatically open and close views based on selection. This lets you select in which perspectives this feature should be enabled. Normally you would only want to enable this in the Debug perspective.

Click **Step Over** once and then click **Step Into** to get into the `read_file()` function. Note that a second stack frame appears in the Debug view and `sort_utils.c` is opened in the Editor. At this point it would be worth taking a closer look at the four tabbed views in the upper right of the workbench.

Table 4.1: Debug Tool Bar Buttons

Button	Name	Function
	Remove all Terminated Launches	Clear all terminated processes in the Debug view.
	Restart	Start a new debug session for the selected process.
	Resume	Resume execution of the currently suspended debug target.
	Suspend	Halt execution of the currently selected thread in the debug target.
	Terminate	End the selected debug session and/or process. Behavior depends on the selected item's type.
	Disconnect	Detach the debugger from the selected process.
	Step Into	If the execution point (program counter) is on a line that includes a function call, step into the function and stop at its first line.
	Step Over	Step over any called functions in the current source line and stop at the next line in the current function.
	Step Return	Execute to the end of the current function and stop at the next line of the caller.
	Drop to Frame	Re-enter the selected stack frame.
	Instruction Stepping Mode	Activate instruction stepping mode to follow execution one instruction at a time.
	Use Step Filters	Activate step filters in the Debug view.

4.6.2 Variables View

When a stack frame is selected in the Debug view, the Variables view displays all the local variables in that frame. Figure 4.8 shows the Variables view for the `read_file()` stack frame. The two variables visible are both pointers. Clicking the white arrow to the left of the name de-references the pointer and displays the corresponding value. For string variables, the full string is displayed in the lower window.

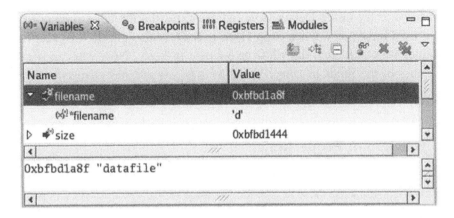

Figure 4.8: Variables view.

Select the `main()` stack frame in the Debug view and note that the Variables view changes to show the local variables in `main()`. If anything other than a stack frame is selected, the Variables view goes blank. Remember, you can also view the value of a variable simply by rolling the cursor over it.

4.6.3 Breakpoints View

To debug the problems in record_sort, you'll probably want to set one or more breakpoints in the program and watch what happens. Select the Breakpoints view, which is currently empty because we haven't set any breakpoints.

Let's set a breakpoint at line 34 in `sort_utils.c`. That's the beginning of an if statement in `read_file()`. Right-click in the marker bar at line 34 and select **Toggle Breakpoint**. A green circle appears to indicate that an enabled breakpoint is set at this location. The check mark indicates that the breakpoint was successfully installed. By the way, there's an editor preference to display line numbers in the editor window.

Select **Window -> Preferences -> General -> Editors -> Text Editors**. The line number display can also be toggled from the marker bar context menu.

A new entry appears in the Breakpoints view with the full path to the breakpoint. The check box shows that the breakpoint is enabled. Click on the check box to disable the breakpoint and note that the circle in the marker bar changes to white. Disabled breakpoints are ignored by the debugger. Click the check box again to re-enable it.

Click **Resume** in the Debug view tool bar and the program proceeds to the breakpoint. The Thread [0] entry in the Debug view indicates that the thread is suspended because it hit a breakpoint. Click **Step Over** to load `temp` with the first record. Select the Variables view and click the white arrow next to `temp`. Now you can see the current values of the fields in `temp`. Variables whose value has changed since the last time the thread was suspended are highlighted in yellow.

4.6.4   Breakpoint Properties

Breakpoints have some interesting properties that extend their flexibility and usefulness. Right-click on the entry in the Breakpoint view and select **Properties** to bring up the dialog in Figure 4.9.

Actions

Actions can be attached to a breakpoint such that when it is hit, the attached actions are performed. CDT offers four classes of pre-defined actions:

- **Play Sound**: Play a selected sound file when the breakpoint is hit. Maybe the breakpoint only happens every half hour or so. You can go off and do something else and when you hear the appropriate beep, you can go back and see what happened. Sound files can be .wav, .mid, .au, or .aiff.

- **Log Message**: Output a message to the console.

- **Run External Tool**: Execute a program that has been configured in Eclipse as an external tool. For example, the program might be running on a remote device. You could configure the breakpoint to send an email or SMS to your desktop

- **Resume**: Automatically resume the program after a specified delay. Again, if the program is running remotely, this is probably the only way to keep it running after a breakpoint.

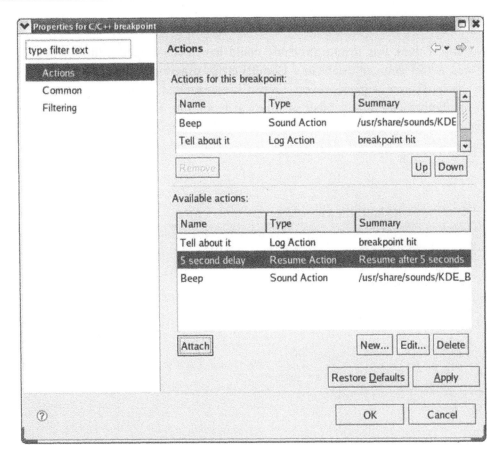

Figure 4.9: Breakpoint properties—actions.

From the **Actions** dialog shown in Figure 4.9 you can create and edit actions using the four classes. Then you can *attach* one or more of these actions to the selected breakpoint. The actions are executed in the order that they appear in the **Actions for this breakpoint list**. Actions can be moved up and down in the list.

Try it out. Create a log action and a resume action and, if you're so inclined, a sound action. Attach them to the breakpoint at line 34 and watch what happens. Be sure you click **Apply** after setting up the action list.

Common

Personally, I think this feature probably could have been named a little better. The idea is that you can establish a **Condition** for triggering a breakpoint rather than have the breakpoint triggered every time it's encountered. You can also specify, with the **Ignore count**, a number of times to ignore the breakpoint before triggering it. Figure 4.10 shows an example where the breakpoint will be triggered when the for loop variable "i" equals 9. The same behavior could be obtained by setting the Ignore count to 9.

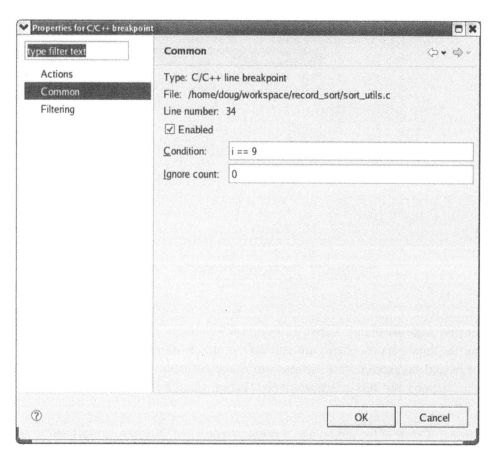

Figure 4.10: Breakpoint properties—Common.

Filtering

Filtering lets you restrict a breakpoint to some subset of threads. Suppose, for example, you have several threads running the same code. You might set a breakpoint in the code, but you're really only interested in stopping one particular thread. You can restrict the breakpoint to just that thread. We'll look at multi-threaded debugging in the next chapter.

At this point it might be useful to let the program execute to line 42, nrecs += 10. Disable the breakpoint. Select line 42 in the Editor window and right-click for the context menu in the text area, not the marker bar. It's worth noting that the context menus for the marker bar and text area are substantially different. Select **Run to Line**. The program executes to the selected line. This is simply an alternate form of execution control.

4.6.5   Other Views

There are two other views that are included in the upper right tab by default. They are the Registers and Modules views. The Registers view shows you the contents of the processor registers. This is only really useful if you're debugging assembly language code. At "C-level" it doesn't tell you much.

The Modules view shows what modules are loaded to create the complete program. Typically this consists of your own executable, plus some collection of shared libraries. Information provided for the executable includes a much more extensive version of the information found in the Outline view.

Memory View

There's one more debug-oriented view that shows up by default in the bottom-tabbed window of the Debug perspective. The Memory view lets you monitor and modify process memory. Memory is organized as a list of *memory monitors*, where each monitor represents a section of memory specified by a base address. Each memory monitor can be displayed in one or more of four predefined formats known as *memory renderings*. The predefined renderings are hexadecimal (default), ASCII, signed integer, and unsigned integer.

Figure 4.11 shows a memory monitor of the area allocated for temp just after the first fscanf() call in read_file(). The Memory view is split into two panes, one that lists the currently active monitors and another that displays the renderings for the selected monitor. A monitor may have multiple renderings enabled, and these will be tabbed in the renderings pane.

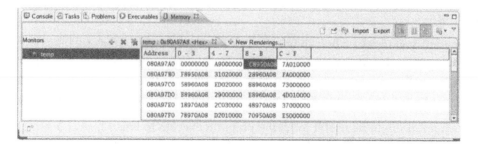

Figure 4.11: Memory view.

The first four bytes hold a pointer to another area allocated for the name. This also happens to be visible. Remember that the x86 is a "little endian" machine. Consequently, when memory is displayed as shown here, most entries appear to be "backwards."

Each of the panes has a small, fairly intuitive set of buttons for managing the pane. These allow you to either add or remove monitors or renderings, and in the case of monitors, remove all of them.

4.6.6   Finish Debugging

With this background on the principal debugging features of Eclipse, you should be able to find the two runtime errors that have been inserted in sort_utils.c. Good luck.

4.7   Linking Projects

Eclipse allows projects to refer to, and to depend on, other projects. Consider, for example, that you have a family of similar application projects that all make use of a common library. We can have each of the application projects refer to the library project so that any time a change is made in the library, the applications are rebuilt. The library project becomes a dependency of the applications.

To illustrate this idea, we'll take the five functions in sort_utils.c and turn them into a static library that we'll link with record_sort.c. Switch back to the C/C++ perspective, select **File –> New**... **C Project** and select **Static Library** as the project type. Call it "sort." Click **Finish**. Right-click on the sort project in Project Explorer and select **Import –> General –> File System**. Browse to EclipseSamples/sort/. Click **Select All** and **Finish**. Five C files and one header file are imported into the project. Build the project. The result is libsort.a.

Now go back to the record_sort project and select **Properties** from the Project Explorer context menu. Select **Project References** and check sort, which may be the only entry (Figure 4.12). Delete sort_utils.c from the file list.

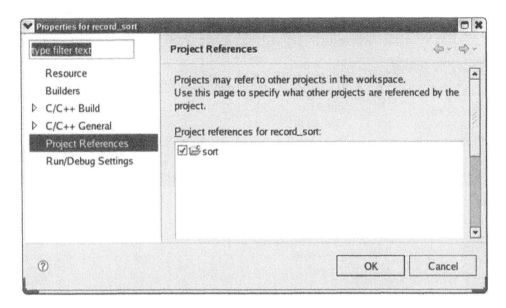

Figure 4.12: Project References.

Now expand the **C/C++ General** entry in the left-hand pane and select **Paths and symbols**. Click the References tab and check sort. This will expand the sort entry and you'll see that the Active build configuration is selected. Select the Library paths tab and you'll see that /sort/Debug has been added (assuming that Debug is the active configuration for sort).

Expand the **C/C++ Build** entry and select **Settings**. Then select **GCC C Linker –> General** and check **No shared libraries**. We'll be using static libraries in this example. Now select **Libraries** (Figure 4.13). The **Library search path** lists the path to the sort project.

Click the **Add** icon in the **Libraries** section and enter "sort."[1] In the context menu for record_sort in the Project Explorer, clean the project. This will rebuild it using the sort library. To prove that record_sort is now dependent on sort, clean the sort project.

---

[1] I find it a little odd that Eclipse can't automatically insert the sort library here, since it apparently figured out that sort is a library project.   ●

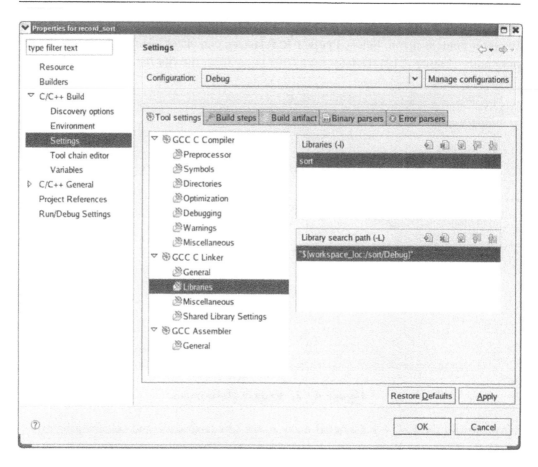

Figure 4.13: Library selection.

This causes it to be rebuilt with a later timestamp on the library than that of the record_sort executable. Now build record_sort. It will be relinked with the new library.

4.8 Refactoring

Refactoring is generally described as a process of restructuring code for the purpose of readability, performance improvements, reuse, or simply for the "regular evolutionary quest for elegance."[2] Most programmers engage in a process of refactoring without

---

[2] http://wiki.eclipse.org/FAQ_How_do_I_support_refactoring_for_my_own_language?

consciously realizing it. As a project progresses we often realize that the approach we started with isn't quite working and it's time to do some prudent restructuring and reorganization.

Eclipse supports an automated approach to refactoring that makes sure that changes are propagated properly throughout a project. The nature of refactoring support is somewhat language-dependent. Eclipse offers extensive refactoring support for Java involving operations on classes, interfaces, and methods. For the moment, anyway, C/C++ refactoring is limited to renaming. While this may seem trivial, propagating a name change accurately throughout a large project can be burdensome if done by hand.

We'll use the record_sort project to illustrate the refactoring support in CDT. Go back to the C/C++ perspective if you're not already there and open record_sort.c if it isn't already. In line 23 select read_file. Now from the main menu, select **Refactor –> Rename**... This brings up the dialog box in Figure 4.14. Let's change the

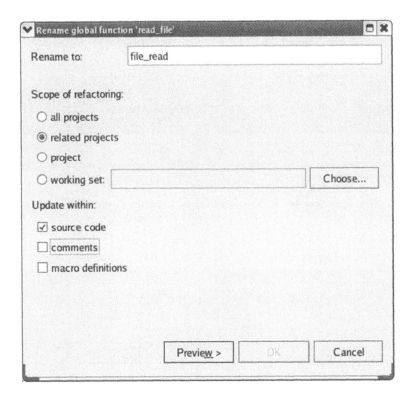

Figure 4.14: Renaming dialog.

name `read_file` to `file_read` by changing it in the **Rename to**: field. The default scope is related projects. This is what we want, because we do in fact have a related project that includes this symbol name. The rename dialog is also accessible from the editor's context menu.

With the **Rename to**: field modified, the **Preview** button is now active. Click it to bring up the dialog in Figure 4.15. This shows all of the places in both projects where `read_file` can be changed to `file_read` and offers the option to selectively change any or all of them. Click **OK** and `read_file` will be magically changed to `file_read` throughout both projects.

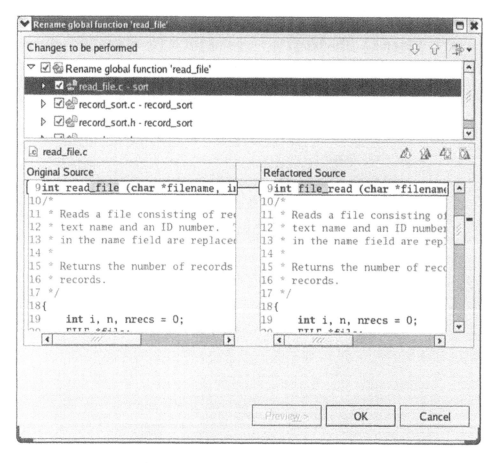

Figure 4.15: Rename change dialog.

Clearly, this example is trivial, but in a large project involving hundreds, perhaps thousands of files, the ability to automatically rename a symbol throughout the project can be quite powerful.

Summary

This chapter has introduced the basic concepts and operation of the C/C++ Development Tools, CDT. We saw how to create a project and looked at the many tools the C Editor offers to ease the task of coding. We examined many of the views in the C/C++ Perspective in detail, to see how they contribute to the development process.

After building the project, we looked at the Debug perspective in detail, with particular emphasis on the many capabilities of breakpoints. We also looked at other views in the Debug perspective to see how they help you gain insight into what's happening in your program.

Finally, we explored a couple of nifty features of Eclipse that can make life easier for developers. Projects can refer to, and can be dependent on, other projects. When changes are made to the referenced project, the dependent project will be rebuilt. Refactoring offers a way to automatically rename a symbol throughout a project or a set of related projects.

The next chapter goes deeper into CDT. In particular, we'll look at how to connect a remote embedded target and debug a program on it.

Resources

Fowler, Martin, et al. 2002. *Refactoring: Improving the design of existing code.* Addison-Wesley.

This book takes a task that many programmers do intuitively and recasts it in a formal methodology.

CHAPTER 5

Eclipse CDT—Digging Deeper

The last chapter covered the basics of the C/C++ Developer Tools: basically how to create, build, debug, and run projects. In this chapter we'll get into more advanced concepts, such as supplying your own makefile and connecting CDT to a remote embedded target.

The project we'll be using in this chapter is called "thermostat." It simulates the operation of a thermostat controlling a cooler. When the indicated temperature rises above the setpoint, the cooler turns on. If the temperature exceeds a limit value, an alarm indicator flashes. The thermostat can also be built for an ARM-based embedded target board.

A separate program provides a means to set the indicated temperature and to observe the cooler and alarm outputs.

5.1 User-Supplied Makefiles

In the last chapter we created a project where Eclipse supplied the Makefile. But there are situations where it makes more sense to create the Makefile yourself and import it into the project along with the source files. This is also useful for bringing "legacy" projects into Eclipse that already have a makefile.

In the C/C++ perspective, create a new C project (**File –> New –> C Project**) but this time select **Makefile Project** as shown in Figure 5.1. Call the project "thermostat." Eclipse creates an empty project with an Includes entry and attempts to build the "all" target. Since the project is empty, there is no "all" target and the build fails.

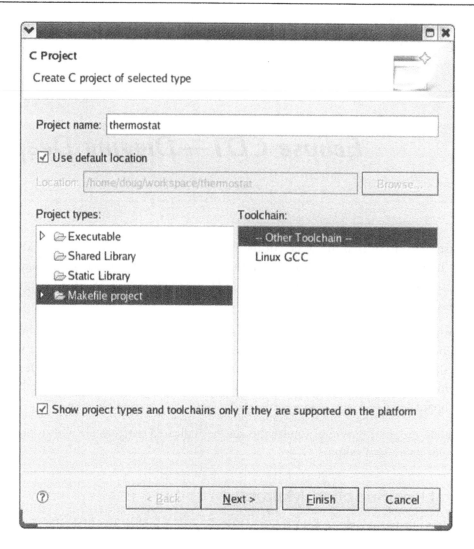

Figure 5.1: Makefile Project.

Right-click **thermostat** in the Project Explorer and select **Import** –> **General** –> **File System**. Browse to the EclipseSamples/thermostat directory and click **OK**. Now click **Select All** and **Finish** as shown in Figure 5.2 to import the entire collection of source files into your thermostat project. Eclipse again attempts to build the "all" target and oddly, it fails even though the Makefile does in fact include an "all' target.

Right-click **thermostat** and select **Build Configurations** –> **Build** –> **All**. This time Eclipse builds an executable target called thermostat_s ("s" for simulation).

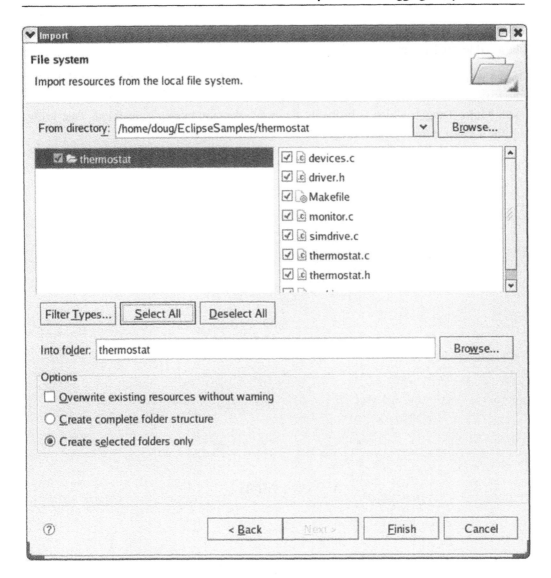

Figure 5.2: Import File System dialog.

5.1.1 Make Targets

Open the Makefile in the Editor window. You'll notice that it builds three different targets, one of which is intended to run on an embedded computer board.

By default, the only thing CDT knows how to make is "all." In this case, "all" can be one of two targets depending on the environment variable TARGET. Then there's an additional build target for the simulated devices program. We have to tell Eclipse about these additional targets.

In the right-hand tabbed window, select the **Make Target** view and right click on the thermostat entry, which brings up a context menu as shown in Figure 5.3. Select **Add Make Target**, which is probably the only selection currently enabled. That brings up the **Create a new Make target** dialog shown in Figure 5.4.

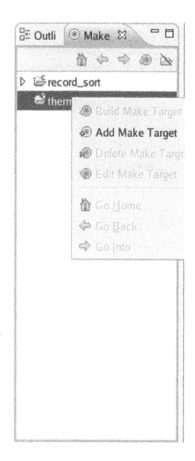

Figure 5.3: Make Target view.

Fill it out as shown in Figure 5.4 and click **Create**. There is now a new entry under thermostat in the Make Target view named "devices." Right-click that entry and select **Build Make Target**. The devices program builds.

Like most operations in Eclipse, there are several ways to create a Make target. With the thermostat project name highlighted in either the **Make Target** or **Project Explorer** views, select **Project** –> **Make Target** –> **Create** ... Enter the **Target Name** as "target" and the **Make Target** as "all." Uncheck the **Use default** box and add "TARGET=1" after make in the **Build command** box.

This target will only build if you happen to have an ARM-Linux tool chain. More on that later in the chapter.

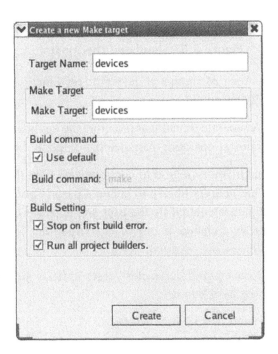

Figure 5.4: Create make target dialog.

5.2   Thermostat Internals

Let's take a closer look at how the thermostat program is organized before we run it under the debugger. As shown in Figure 5.5, thermostat consists of three modules:

- `thermostat.c`—The `main()` function that implements the thermostat state machine and outputs the current temperature to `stdout`. `main()` takes a run-time argument that is the delay between samples in seconds. The default is two seconds.

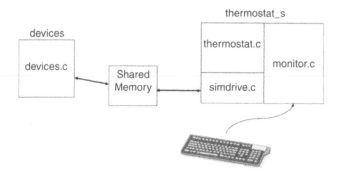

Figure 5.5: Simulated thermostat.

- `simdrive.c`—A set of simulated device driver functions for an A to D converter and digital output.

- `monitor.c`—A separate thread that monitors the keyboard for operator input to change the thermostat operating parameters—setpoint, limit, and deadband.

`simdrive` uses a shared memory region to communicate with the simulated `devices` program. devices uses the ncurses library to create a pseudo-graphical user interface. It just reads A/D input values entered by the user and depends on a signal sent by `thermostat_s` to indicate a change in one of the output bits. Entering a non-numeric value for the A/D terminates the program.

The command syntax for changing parameter values is fairly simple. It's just a single lower case letter followed by a number:

- s—setpoint

- l—limit

- d—deadband

The monitor thread just sets the new value into the corresponding global variable. Although it's probably not strictly necessary in this case, access to the global variables is protected by a mutex because that's the right thing to do.

Have a look through the four source files to see how it all hangs together.

5.2.1 Running the Simulation

The devices program needs to be started first. We'll run it from a shell window outside of Eclipse. In a shell window, cd to ~/workspace/thermostat (note that "~" is a shortcut for your home directory). Both executables, `thermostat_s` and `devices`, are in this directory. Enter `./devices` to start the devices program.

Back in Eclipse, right-click `thermostat_s` under the thermostat project in the Project Explorer view and select **Debug As –> Debug Configurations** ... to create a new debug configuration for this project. Name it whatever you'd like and select the thermostat project. **Search Project** doesn't work for finding an application in a Makefile project, so you have to browse to the project directory in your workspace and select the file, `thermostat_s`. If you choose, enter a number in the Arguments tab for the program's loop delay time in seconds. Click **Apply** and then **Debug**.

As before, Eclipse brings up the Debug perspective with the program stopped at the first executable line of `main()`.

5.3   Debugging Multi-Threaded Programs

The thermostat program uses POSIX threads (pthreads) to create two independent threads of execution: the thermostat state machine, and the monitor that looks for parameter changes. Fortunately, Eclipse and gdb are very good at handling multi-threaded programs.

Set a breakpoint at the call to `createThread()` and resume the program. Step into `createThread()` and step until the `pthread_create()` function executes. The Debug view now shows that a second thread has been created (Figure 5.6). Thread [1] is suspended in `createThread()` as we expect.

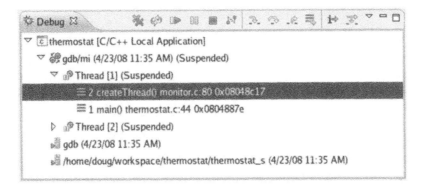

Figure 5.6: Multi-threaded debug.

Thread [2] is also suspended. Expand its entry in the Debug view and you'll see that it is suspended in the `clone()` function. `clone()` is a Linux system service, that like `fork()`, creates a child process but offers finer control over what gets shared between parent and child. Thus it is useful for creating kernel-schedulable threads that share their parent process's global data space.

At this point we would probably want to gain control of the monitor thread to watch its execution. `monitor.c` should already be open in an Editor window since we've been stepping through a function in that file. Scroll back up to find the `monitor()` function that executes Thread [2]. How do we know that that's the right function? Well, `monitor` is the *start_routine* argument to `pthread_create()`.

Set a breakpoint on the line following the call to `fgets()`. `fgets()` doesn't return until you enter a string on `stdin`. Click the **Step Return** button to get back to `main()`. Expand Thread [2]'s entry in the Debug view and you'll see that it's now way down deep somewhere in the kernel having called `fgets()` from `monitor()`.

At this point you might want to step through `initDigIO()` and `initAD()` just to see what they do. When you're finished, click the **Resume** button to let the program execute. The Debug view shows that both threads are Running and sample data starts appearing in the Console view.

Enter a new A/D value in devices and watch it show up in the thermostat output. In the same Console view, enter a parameter command, say "s 44," to change the setpoint. Note that the Console view must have the focus to type something into it. This causes both threads to suspend with Thread [2] at the breakpoint, as we would expect. Thread [1] is deep inside the kernel having called the `sleep()` system service. You can now step through the execution of `monitor()` as it parses the command you just entered.

The next thing we might want to do is trace the execution of Thread [1] as it executes a state change. `thermostat.c` should still be open in an Editor window, so select that tab. Set a breakpoint on the `switch (state)` line. Now in the Breakpoints view, open the Properties page for the breakpoint you just created and set the Condition as "value > setpoint + deadband." Incidentally, you can simply copy and paste that expression from `thermostat.c` rather than having to type it.

Let the program resume and enter a new A/D value that is in fact greater than the setpoint you just entered plus the default deadband of 1. The condition for our new breakpoint is satisfied, and so the program suspends. We can now step through and verify that a state transition occurs and the "cooler" gets turned on.

Resume the program, but note that the breakpoint will be triggered again on the next pass through the loop because the condition is still true. You might want to change the condition to observe the transition to the LIMIT state.

This then is the general strategy for debugging multi-threaded applications. Strategically place breakpoints in the threads of interest and watch them interact. But what if you guess wrong and none of your breakpoints are triggered, or the breakpoints aren't triggered because the program isn't behaving as you expect?

In this case the **Suspend** button lets you regain control of the program. **Suspend** sends a SIGINT signal to the currently selected thread to suspend the program. Chances are the thread is not executing your source code but is somewhere inside the kernel with no symbol information available. This brings up a new Editor window with the notation "No source available for """ and a **View Disassembly** button.

You can click **View Disassembly** to see the actual machine code being executed when the signal was received, but it's probably not terribly enlightening. The most useful thing to do is select the stack frame for your thread function to show where it is executing. Set a breakpoint immediately after that line and you'll regain control in your thread function.

5.3.1 The Signals View

The suspend functionality raises the interesting question of how gdb handles signals in general. Normally, the SIGINT signal is generated by typing **Ctrl-C**, which causes the receiving process to be terminated if it doesn't handle the signal itself.

The thermostat program registers a signal handler for SIGINT to terminate the program gracefully. Among other things, in a multi-threaded environment, it's important that threads terminate in the correct order. That's what the function `terminateThread()` does.

But by default gdb intercepts the signal and uses it to suspend the program. Click **Window –> Show View –> Signals**. This displays the Signals view in the upper right tabbed window (Figure 5.7). This shows you the signals defined on the selected target and how the debugger handles each one.

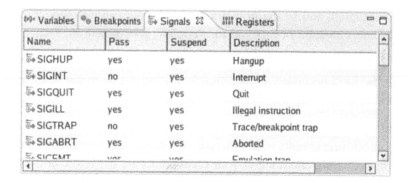

Figure 5.7: Signals view.

SIGINT is currently configured to suspend the program, which is the behavior we just saw, and not to pass the signal on to the program. Right-click the SIGINT entry and select **Signal Properties**. This brings up a dialog with two check boxes:

- Pass this signal to the program.

- Suspend the program when this signal happens.

If both boxes are checked, the program is suspended before the signal is passed to the program. Resuming the program then allows the signal handler to be executed. With the Pass box checked and the Suspend box unchecked, the signal handler is executed first, and then oddly enough, the debugger catches the signal later. Try it with thermostat and you'll see that Thread [1] is suspended in the `pthread_cancel()` call inside `terminateThread()`.

5.4   Working With Embedded Target Hardware

While the simulated thermostat is a useful way to illustrate many Eclipse concepts, and it can help you start testing software before the target hardware is ready, eventually you'll need to test your program on real hardware. Eclipse has facilities to support that process.

5.4.1   System Requirements

The following hardware and software components are required in order to work through the example in this section:

- Target embedded computer board with network and serial ports. The example here is based on a board available from Intellimetrix. See Appendix B,

"The Embedded Linux Learning Kit," for details on this board and how to obtain it. A nice feature of this board is that it includes a temperature sensor for the thermostat example.

- Serial and network ports on the host workstation.

- GCC cross tool chain for the target board.

- NFS server running on the host machine.

- minicom terminal emulator for communicating with the target board's serial port.

- TFTP server is useful but not required.

All of the software components with the exception of the cross tool chain should be part of any decent Linux distribution. The tool chain should be available from the target board vendor.

The other approach to a target environment for the purpose of this chapter is to make use of an old 486 box that's sitting in the closet or serving as a doorstop. Put a network card in the box and install Linux on it. Redirect `stdin`, `stdout`, and `stderr` to a serial port. In this case you don't need a cross compiler. The one you've been using is quite sufficient.

5.4.2 The Cross-Development Environment

Very often a target board stores its root file system in a non-volatile or "semi-volatile" medium such as flash memory. This can make it difficult and time-consuming to load executable images onto the target for testing. This is especially true in early stages of testing, where the edit-compile-debug cycle turns rapidly.

The development paradigm we'll follow here makes use of the Network File System (NFS) to remotely mount part (or even all) of the target's file system on the host workstation. Then we simply put our target executables in a location on the host that's visible to the target, which then loads the image off the host and executes it.

Figure 5.8 illustrates the process graphically. The `/home` directory on the target is remotely mounted through NFS to ~/`workspace`, the default Eclipse workspace. Now all the project subdirectories of ~/`workspace` on the host show up as subdirectories of `/home` on the target. The serial port serves as the target shell's console communicating with the minicom terminal emulator running on the host. Technically, you could telnet to the target shell and get by without the serial port, but in many cases a serial port is still necessary to communicate with the target's boot loader.

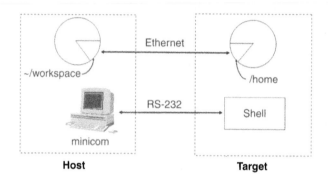

Figure 5.8: Cross-development paradigm.

5.4.3   Host Configuration

There are three aspects to configuring the host workstation for target development:

1. Install the GNU cross tool chain. The supplier of the tool chain should provide instructions and/or scripts for this. If your target is a PC, this step isn't required.

2. Configure the terminal emulator, minicom.

3. Configure networking.

Configure Minicom

minicom is a fairly simple Linux application that emulates a dumb RS-232 terminal through a serial port. This is what we'll use to communicate with the Linux system running on the target board.

There are a number of minicom configuration options that we need to change to facilitate communication with the target.

In a shell window as root user, enter the command minicom –s. If you're running minicom for the first time you may see the following warning message:

```
WARNING: Configuration file not found. Using defaults
```

You will be presented with a configuration menu. Select Serial port setup (Figure 5.9). By default, `minicom` communicates through the modem device, `/dev/modem`. We need to change that to talk directly to one of the PC's serial ports. Type "A" and replace the word "modem" with either "ttyS0" or "ttyS1", where ttyS0 represents serial port COM1 and ttyS1 represents COM2. However, if your host only has USB ports and you're using a USB-to-serial converter, the correct device is most likely "ttyUSB0."

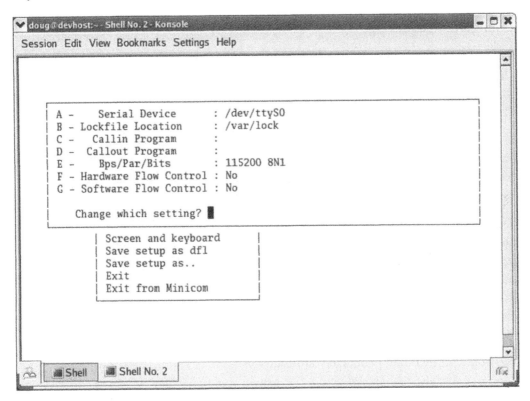

Figure 5.9: minicom serial port setup.

You'll need to change the data rate (bps) to match your target board. Chances are you won't want either hardware or software flow control.

Type **Enter** to exit Serial port setup and then select Modem and dialing. Here we want to delete the modem's Init string and Reset string since they're just in the way on a direct serial connection. Type "A" and backspace through the entire Init string. Type "B" and do the same to the Reset string.

Type **Enter** to exit Modem and dialing. In the Screen and keyboard menu, you may need to change the backspace behavior for shell line editing to work correctly. Finally, select **Save setup as dfl** to save the configuration.

You will probably have to change the permissions on the device node for the selected serial port to allow the group and world to read and write the device. And of course, you must be root user to do this.

Configure Networking

Your workstation is probably configured to get a network address via DHCP (Dynamic Host Configuration Protocol). But in this case, to keep things simple, we're going to specify fixed IP addresses for both ends. This is particularly useful if you choose to directly connect the workstation to the target using an Ethernet crossover cable.

If you're using KDE, there's a nice graphical menu for changing network parameters. How you get to that menu from the Start menu varies depending on the distribution. The two variants I've seen so far are **System –> Network Configuration** and **System Settings –> Network**. You'll be asked for the root password.

The actual layout of the dialog boxes will vary by distribution. You should find a **Devices** tab and then be able to edit the entry for your Ethernet device, which is probably eth0. The edit dialog allows you to select between "Automatically obtain IP address settings" and "Statically set IP addresses." Select the latter and assign an address. I usually assign 192.168.1.2 to the workstation. The dialog from Red Hat Enterprise Linux 4 is shown in Figure 5.10.

Alternatively, you can just go in and directly edit the network device parameters file. Network configuration parameters are found in /etc/sysconfig/network-scripts/ where you should find a file named something like ifcfg-eth0 that contains the parameters for network adapter 0. You might want to make a copy of this file and name it dhcp-ifcfg-eth0. That way you'll have a DHCP configuration file for future use if needed. Now open the original file with an editor (as root user of course). It should look something like Box 5.1a. Delete the line BOOTPROTO=dhcp and add the four new lines shown in Box 5.1b.

We'll use NFS (Network File System) to download executable images to the target. That means we have to "export" one or more directories on the workstation that the target can mount on its file system. Exported directories are specified in the

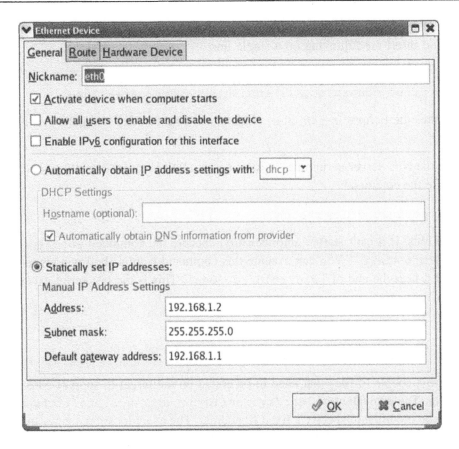

Figure 5.10: Ethernet configuration dialog.

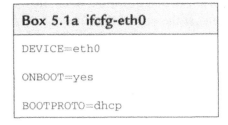

Box 5.1a ifcfg-eth0

```
DEVICE=eth0

ONBOOT=yes

BOOTPROTO=dhcp
```

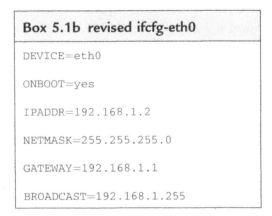

Box 5.1b revised ifcfg-eth0

```
DEVICE=eth0

ONBOOT=yes

IPADDR=192.168.1.2

NETMASK=255.255.255.0

GATEWAY=192.168.1.1

BROADCAST=192.168.1.255
```

file `/etc/exports`. Initially this file is present but empty. As root user, open it with an editor and insert the following on a single line:

```
/home/<your_home_name>/workspace
*(rw,no_root_squash,no_all_squash,sync,nosubtree_check)
```

This makes the Eclipse `workspace/` directory visible to other computers on the network.

Be sure the NFS server is running. In most cases it will be automatically started at boot time. Use the command:

```
/etc/rc.d/init.d/nfs status
```

to check this. If it isn't started automatically, execute the command `/etc/rc.d/ init.d/nfs start.`[1] You can execute this command from a shell window or, better yet, add it near the end of `/etc/rc.d/rc.local`. This is the last script executed at boot up.

5.4.4   Target Configuration

There are a couple of files that need to be edited on the target to support NFS file mounting. The equivalent of `/etc/sysconfig/network-scripts/ifcfg-eth0` needs to be modified to specify a fixed IP address. The exact location, name, and layout of this file will vary from board to board. On the Intellimetrix board it's `/etc/ network/interfaces`. I usually assign 192.168.1.50 to the target board.

The other file that needs modification is the last script executed by the init process when the system boots. Again, this will vary from board to board. On the Intellimetrix board it's `/etc/init.d/rcS`. Add the following line at the end of that file:

```
/bin/mount -o nolock 192.168.1.2:/home/<your_home_name>/workspace
/home
```

On a PC target `/etc/ rc.d/rc.local` is a good place to add this.

---

[1] This is the location of the nfs script on a Red Hat or Fedora distribution. Other distributions may locate it somewhere else.

This causes the `workspace/` sub-directory under your home directory to appear as `/home` on the target using the Network File System.

In order to debug on the target, you'll need a program called `gdbserver` compiled for the target and loaded on the target's file system, preferably someplace visible from the `PATH` environment variable. `gdbserver` runs the program under test and communicates over the network with GDB running under Eclipse on the host.

5.4.5 Creating a Target Eclipse Project

Even though our thermostat project included a make target for the target version of the thermostat, we'll create a new project to illustrate some additional features of Eclipse. Create a new C executable project and call it "target." From the thermostat project, select the following files using the Shift and Ctrl keys in the same way you do when selecting multiple files in Windows or graphical Linux environments:

- `AT91RM9200.h`
- `driver.h`
- `thermostat.h`
- `monitor.c`
- `thermostat.c`
- `trgdrive.c`

Right-click and select **Copy**. Click on the new target project, right-click and select **Paste**. We now have all the files we need for the project. But remember that we don't want to build this project for the host, but rather for the target. This requires configuring the project to use a different GNU tool chain.

The file `trgdrive.c` provides a set of device driver functions for the Intellimetrix ARM9 board. This is not a real Linux "device driver," but rather accesses memory mapped I/O directly from user space. If you're using some other target board, you'll need to modify the functions in `trgdrive.c` accordingly.

If your target is a PC, you can skip to the next section on debugging on the target. You don't need to select a different compiler.

Right-click on the project name and select **Properties**. Expand the **C/C++ Build** entry and select **Settings**. This brings up the dialog shown in Figure 5.11. The first tab, Tool settings, lets you specify which C compiler, C linker, and assembler to use. By default, the compiler and linker are just gcc. This is the standard name for a host GNU C compiler.

By convention, cross compilers are given a prefix that identifies the architecture and the operating system on which the compiled program will run. On my system, the ARM

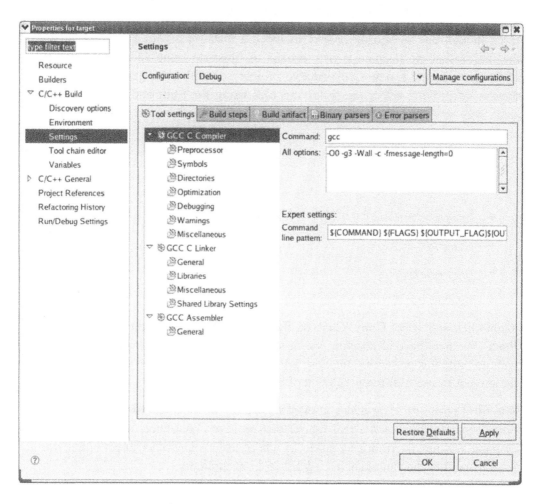

Figure 5.11: Project build settings.

cross compiler is called `arm-linux-gcc` and I've added the path to it to my PATH environment variable. So change the Command name to match your cross compiler. All of the items under **GCC C Compiler** represent categories of compiler options. Take a look through them to see what's there.

Likewise, change the **GCC C Linker** command to match your cross compiler. Here we also have to add a library to the linker command. Select **Libraries**, click the **Add** button, and enter "pthread." This is the library of Posix threads functions. Take a quick look at the categories of linker options.

Finally, change the **GCC Assembler** to match your cross assembler.

Note that we've done all this for the **Debug** configuration. Click the **Configuration** drop-down menu at the top, select **Release**, and make the same changes.

Click **OK** to exit the Properties dialog. Make sure the active build configuration is Debug and build the project. You'll find two new entries under the target project in the Project Explorer view: Binaries and Debug. Debug lists all of the built objects including an executable named "target". Expanding any of the built object entries produces a list of every source file used to build that object. It's not clear to me what purpose that serves.

5.4.6   Debugging on the Target

From a developer's standpoint, there's virtually no difference between debugging on host and debugging on a target board. The only difference is in how Eclipse connects to the debugger. For this you'll need to create a new debug configuration.

With the target project selected in the Project Explorer view, select **Run –> Debug Configurations** ... Click the **New launch configuration** button. A new configuration named target Debug is created, referencing the target project and the Debug/target application. Select the Debugger tab (Figure 5.12).

Open the **Debugger** drop-down menu and select **gdbserver Debugger**. Enter the name of your cross gdb in the **GDB debugger** field. Click the **Connection** tab under Debugger Options. Select **TCP** from the **Type** drop-down and enter the IP address of your target in the **Host name or IP address** field (Figure 5.13). Click **Apply** and then **Close**. GDB is now configured to talk to your target board over the network.

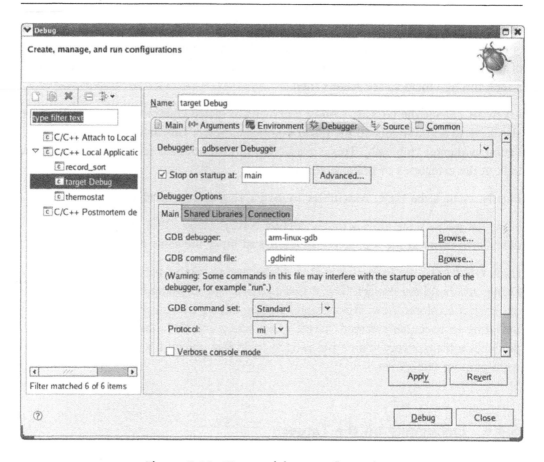

Figure 5.12: Target debug configuration.

But before starting up a debug session, you have to start the application running on the target:

1. Execute `minicom` from a shell window on the host.

2. Power up your target board and boot into Linux.

3. In `minicom`, `cd /home` and verify that it has the same contents as ~/workspace on the host. If not, go back to the section on configuring networking and see if anything isn't right. Try executing the `mount` command manually from the shell.

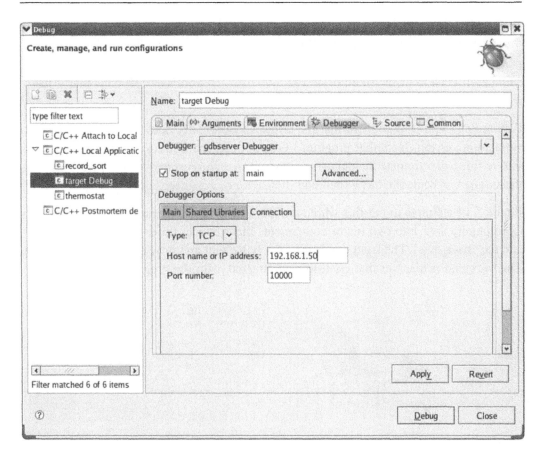

Figure 5.13: Target debug connection.

4. `cd target/Debug`.

5. Execute `gdbserver :10000 target`.

This last command starts `gdbserver` telling it to listen for a connection on TCP/IP port 10000, which is the default port assigned in the Eclipse configuration, and start up the `target` executable. `gdbserver` responds that it created a process for the `target` executable and that it is listening on port 10000. Note, incidentally, that you can use any port number you want as long as both sides use the same number and it doesn't conflict with some other network service. Port numbers below 1024 are reserved for established services and shouldn't be used.

Back in Eclipse, you should be in the C/C++ perspective. Select the target project in the Project Explorer view and execute **Run –> Debug**. Eclipse switches to the Debug perspective and you should see virtually the same thing you did earlier with the simulation version of the thermostat. On the target, gdbserver responds that it has established a debugging session with host 192.168.1.2.

Play around with it to confirm that the target version really does behave like the simulation version. Note that program I/O is through the minicom window. When you're finished terminate the debug session. gdbserver responds that its child process, the target executable, has exited and then gdbserver itself exits.

Figure 5.14 attempts to put this whole process into graphical perspective. GDB itself is conceptually split into two major components that you can think of as the "front end" and the "back end." The front end is effectively a client and provides the user interface. The back end is a server that controls and interacts with the program under test.

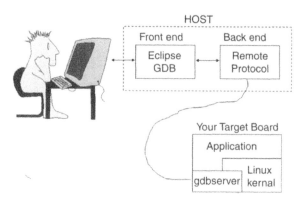

Figure 5.14: GDB architecture.

The back end is architecture-dependent so it knows how to do things like disassemble machine code and set breakpoints. When we debugged the simulation thermostat, the back end simply ran the program under test on the workstation.

Things got a little more complicated when we moved to debugging on a remote target board. Now the back end running on the host has to know that it's not running the program under test locally, but rather becomes a client to gdbserver running the program under test on the target. GDB provides a well-defined protocol for this communication. The primary role of gdbserver is to get and set memory locations and processor registers, and to maintain control over execution of the program under test.

The existence of a well-defined and public protocol for communicating between the GDB back end and gdbserver opens other possibilities for communicating with the target. Specifically, vendors of In Circuit Emulators (ICE) and JTAG boxes can use that protocol to make their products useable from GDB and Eclipse.

Summary

This chapter has delved deeper into the Eclipse C Developers' Toolkit (CDT) to look at issues such as supplying your own Makefiles, cross-compiling for target embedded hardware, and debugging an application on a remote target board. We also explored the notion of using high level simulation to get started with application software testing before any target hardware is available.

The next chapter looks at the Device Software Development Platform (DSDP), another Eclipse project related to embedded software development. In particular, DSDP provides target management capabilities that further simplify the process of connecting to a remote target.

CHAPTER 6

Device Software Development Platform

The Eclipse Device Software Development Platform (DSDP) extends CDT to address specific needs of embedded device software developers. It consists of six subprojects, some of which are farther along than others:

- Target Management (TM)

- Remote System Explorer (RSE)

- Native Application Builder (NAB)

- Embedded Rich Client Platform (eRCP)

- Mobile Tools for the Java Platform (MTJ)

- Tools for Mobile Linux (TmL)

In this chapter, we'll take a closer look at Target Management and the Native Application Builder. The Embedded Rich Client Platform is tied in with plug-in development that we'll explore in the next chapter. Mobile Tools for Java and Tools for Mobile Linux are currently in an "incubation" stage, making them a little difficult to describe accurately and thoroughly.

But before proceeding, we should look at how Eclipse manages updates and extensions.

6.1 Adding on to Eclipse

The beauty of Eclipse is that it can be easily extended through the plug-in mechanism. Currently, your Eclipse installation includes the base platform, CDT, the Rich Client Platform (RCP), and a CVS version control system client. For this chapter we'll install some additional features.

There are a couple of different approaches to adding features to Eclipse. The simplest approach is to download a zip file of one or more plug-ins and just unzip it in your Eclipse directory. Oddly though, plug-ins don't seem to carry any information about what other plug-ins or features they may depend on. Thus, you may install a plug-in only to find that it doesn't work because it's dependent on something else that isn't installed and isn't identified.

Eclipse has addressed this issue with the concept of a *feature*, which bundles together plug-ins that are logically related to perform some useful function. Features are then published to an update site on the Internet from where they can be downloaded by the Eclipse Update Manager. Organizations that create Eclipse plug-ins are encouraged to maintain an update site and publish their plug-ins as features.

Start up Eclipse if it's not already running and select **Help** –> **Software Updates** to bring up the menu in Figure 6.1. The Available Software tab is displayed, listing update

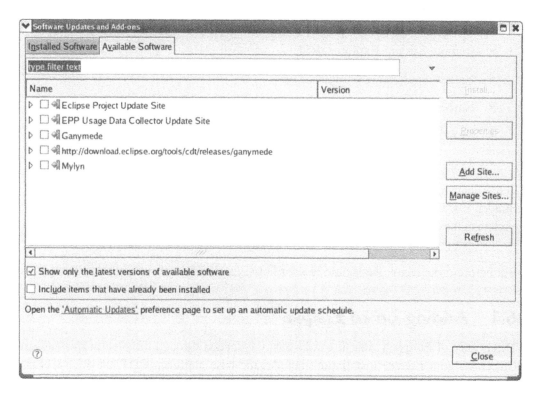

Figure 6.1: Update Manager.

sites that can be searched for new features. You can add remote update sites in the form of URLs or Local sites, such as a CD. An Archive site is a local site packaged as jar or zip files.

Several sites are already available for searching. Expand the **Ganymede** entry. The Update Manager goes to the selected site to discover what new features are available and presents a list of feature categories. Scroll down to and expand **Remote Access and Device Development** (Figure 6.2).

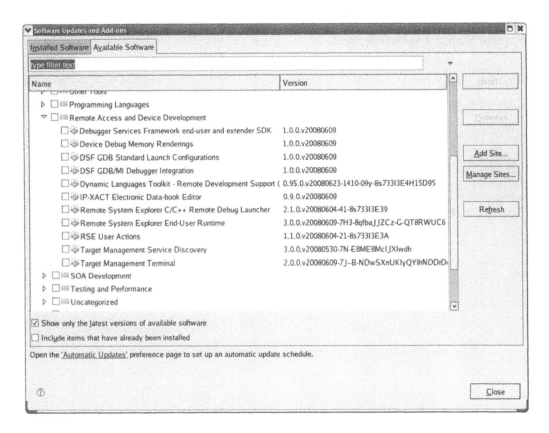

Figure 6.2: Expanded feature category.

Check the box next to **Remote Access and Device Development** to select all 11 of its features. When you click **Install**, the Update Manager automatically resolves any dependencies and downloads them as well. You are asked to review and confirm the items to be installed. Next, you'll be asked to accept the license terms. Click **Finish**.

Install offers the option of running in the background so that you can continue working in Eclipse while the download is in progress. This is a useful feature, since downloads can take a while. When the installation has completed, Eclipse suggests that you restart the system for the changes to take effect. Some changes can be applied without restarting, but generally it's a good idea to restart.

You can get additional details on a feature by selecting it and clicking the **Properties** button. Figure 6.3 is an example of the General Information category for one of the Remote System Explorer features.

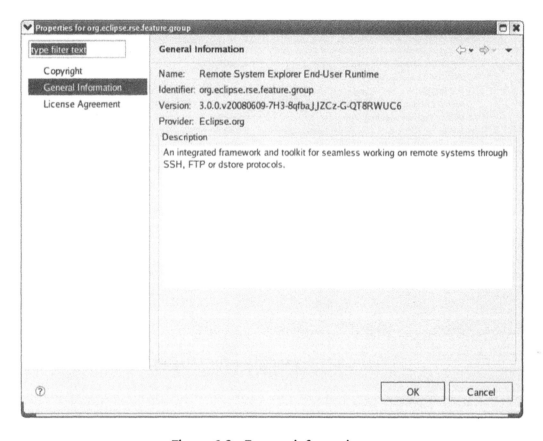

Figure 6.3: Feature information.

6.1.1   Installing Features in External Directories

Normally, features are installed in `eclipse/features` and plug-ins are installed in `eclipse/plugins`. If, for whatever reason, you should need to reinstall Eclipse, all of the additional features and plug-ins you had previously installed would also have to be reinstalled. But if they're installed in an external directory, reinstalling Eclipse itself doesn't affect them.

The previous version of the software update feature offered an option to install features and plug-ins in alternate directories. The current version doesn't seem to have that option. Nevertheless, if you install plug-ins manually, you're free to put them wherever you like.

If you do create an external directory for plug-ins, you have to let Eclipse know about it. On startup, Eclipse looks for a directory `eclipse/links`. In that directory create one or more text files whose names end in .link. Each file contains an entry of the form:

```
path=<path_to_eclipse plug-ins>
```

For example, `path=/usr/local/eclipse-plugins`

6.1.2   Updating Existing Features

The other thing the Update Manager can do is search for and install updates of existing features, including the platform itself. Select **Help** –> **Software Updates** –> **Find and Install** . . . but this time select **Search for updates of the currently installed features**. Click **Finish** and the Update Manager will search for updates to installed features. This can be a lengthy process, so you'll probably want to click **Run in Background** so you can continue working while the update process runs.

The update search can be scheduled to run automatically. Select **Window** –> **Preferences** . . . –> **Install/Update** –> **Automatic Updates** (Figure 6.4). Here you can choose from several update scheduling policies:

- No automatic update
- On every startup
- Every day at a specific time
- On a scheduled day of the week at a specific time

The last two options, of course, require that Eclipse is running at the selected time.

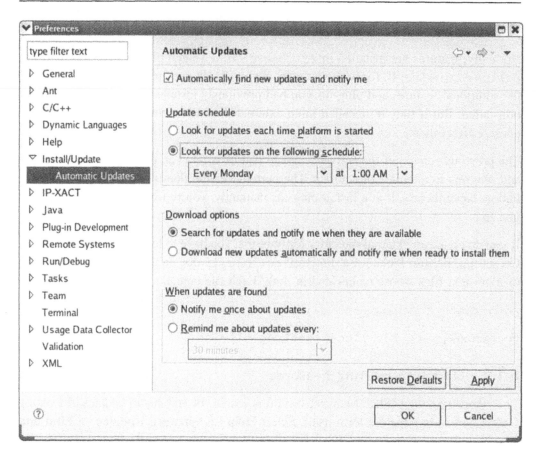

Figure 6.4: Scheduling automatic updates.

6.2   Target Management and the Remote System Explorer (RSE)

The Remote System Explorer (RSE) is a collection of tools that allows you to work with resources such as files and folders on remote systems. The Remote System Explorer perspective allows you to directly manipulate resources on a remote system. The available actions depend on the type of system you're connecting to and the way the resource is recognized.

For a quick tour of some basic RSE capabilities, go to the Remote System Explorer perspective. Currently, there is one item called Local. This, in fact, is the local host.

Expand that item to reveal Local Files and Local Shells. Under Local Files is My Home and Root, which are fairly self-explanatory. Expand My Home, and you'll see the contents of your home directory. Root, of course, is the entire local file system.

Right-click on Local Shells and select **Launch Shell**. This brings up a Remote Shell view in the lower tabbed window. Commands are entered in the small window at the bottom, and the results show up in the larger window. Figure 6.5 shows the result of executing the ls command on my workspace directory. File name completion is done with **CTRL+Space** rather than tab. The Remote Shell view has tool bar icons to clear the results, terminate the shell, and save both command results and command history to files.

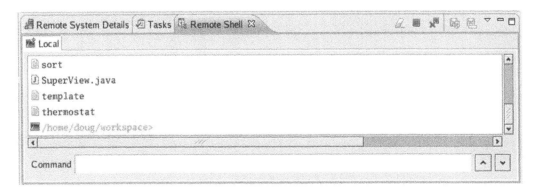

Figure 6.5: Remote Shell.

6.2.1 Connecting to a Remote System

Of course, the Remote System Explorer isn't very interesting until we connect to a remote system. This section assumes you have network access to another computer running Linux.

RSE is a framework that supports plugging in many different communication protocols. By default FTP and SSH (secure shell) are supported along with a more capable RSE-specific protocol called dstore. The latter requires a server running on the remote system. We'll examine both SSH and dstore connections. SSH is easy, and requires fewer resources on the target; dstore has more functionality.

SSH Connection

In the Remote Systems view, click the **Define a connection** icon or right-click on **Local** and select **New –> Connection** ... The first step is to select the Remote System Type. Select SSH Only and click **Next**. The next screen (Figure 6.6) initially displays LOCALHOST as the Host name. Change that either to an IP address or to the name of a computer on your network. The Connection name defaults to the Host name, but you can change it to anything you want.

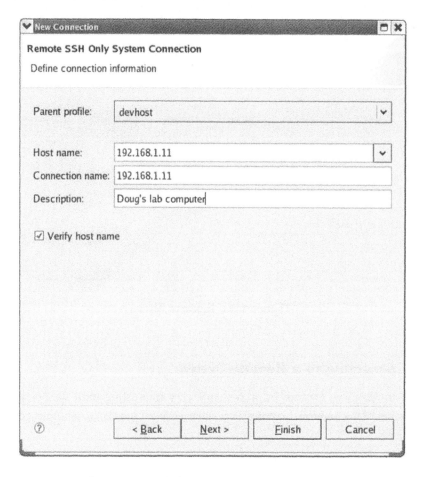

Figure 6.6: Setting up an SSH connection.

Clicking **Next** shows the file services available on the remote machine. Clicking **Next** again shows the available shell services. One final click on **Next** shows the available terminal services. Click **Finish** and the new connection shows up in the Remote Systems view. Incidentally, the SSH daemon must be running on the remote system. Most Linux distributions start it at boot time.

Expand the new connection to reveal entries similar to what we saw with Local. When you expand My Home or Root under Sftp Files, you're required to enter a valid user ID and password for the remote system, which effectively logs you into it. You can now use copy and paste commands to move files between your local host and the remote system. Give it a try.

There's an option to save the user ID and password as defaults, so you don't have to enter them every time.

Right-click on Ssh Shells and select **Launch Shell**. You now have a command shell connected to the remote machine. You can also right-click any directory entry in the remote file system and launch a shell in that directory. This allows for multiple shells to the same remote system. These are opened as tabs in the Remote Shell view.

Right-click a text file in the remote system and select **Open**. The file is opened in the Eclipse editor.

Dstore Connection

The dstore protocol provides more capability than SSH at the expense of requiring a Java-based server running on the remote machine. dstore supports browsing into archive files such as tar, tgz, and zip. It supports remote search without transferring files to your local host. Nevertheless, the requirement for Java means it's probably not the best fit for embedded devices.

Before creating a dstore connection, you must install the dstore server on the remote machine. Download it from the Target Management project download site at http://www.download.eclipse.org/dsdp/tm/downloads. Click on the latest release link, currently 3.0, and in that page scroll down to the DStore Server Runtime. Versions are available for Linux, Windows, other Unix, and Mac OS X (considered experimental). Currently, a server implementation for Windows CE is in "incubation."

Decide where to install the server—on a Linux box /opt is a good place—and create a directory for it, say, /opt/rseserver. Untar the downloaded tar file there.

The dstore server also requires a Java Runtime Environment (JRE) version 1.4 or higher. An IBM, Sun, or equivalent JRE is required. The gcj-based jvm that comes with most Linux distributions doesn't work. If necessary, refer back to Chapter 2, Installation, for information on installing a JRE. The scripts that start the Linux server also require Perl, which is installed by default on most distributions.

There are two ways to start the dstore server on a Linux remote system. As root user you can start a server daemon by executing the Perl script `daemon.pl`. The server daemon listens for connection requests on port 4075 by default. You can change the port number with an optional port argument to the script. In response to a connection request, the daemon spawns a dstore server.

Alternatively, if you don't have root access, or just don't want the daemon running, you can manually start the server with the Perl script `server.pl`. If you don't specify a port, the server picks the first one available and prints out the port number. This is usually 4033. If no connection shows up in about two minutes, the server times out. A manually launched server also terminates when you terminate the connection.

The process is essentially the same for a Windows remote system, except that the scripts are called respectively `daemon.bat` and `server.bat`.

To establish a dstore connection, click the **New Connection** icon and select either Linux or Windows as the remote system type. Specify the host name and a connection name, just as we did for the SSH connection. Click **Next**. In this dialog (Figure 6.7) Launcher Properties specifies how the server is started on the remote system. If you started the daemon on the remote system, then select **Daemon** as the Launcher. If you manually started the server, select **Running**.

Click **Finish**. Right-click on the new entry in the Remote Systems view and click **Connect**.

Personally, I find the SSH Only connection to be simpler and easier to work with. Dstore is useful in connecting to Windows boxes, because standard Windows doesn't include an SSH server.

Creating a Second Connection

There are times when you might want to have two or more connections open to the same remote system. You might want to log on with a different user ID, for example. Or you might want to have both a dstore and an SSH connection open.

With the first connection selected in the Remote Systems view, click the **New Connection** icon or right-click and select **New –> Connection**. The Host name remains

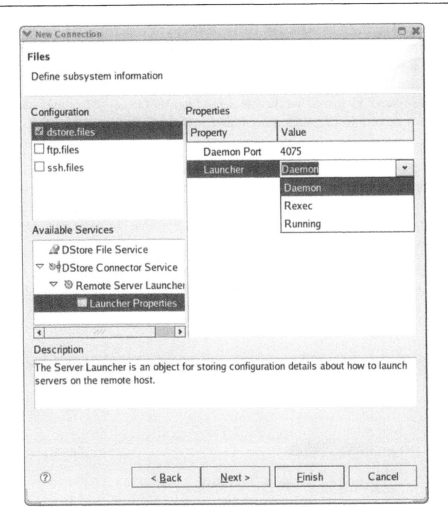

Figure 6.7: Setting launcher properties.

the same. You just need to give it a unique Connection name. Go ahead and create a second connection, because we'll use it to illustrate referencing filter pools later on.

6.2.2 Filters and Filter Pools

The entries My Home and Root under Sftp Files in your first remote connection are known as *filters*. RSE automatically creates these two filters for every

SSH connection. Right-click on **My Home** and select **Properties**. The Filter Information page tells you that this is a Filter Resource and gives its name. It also says that it belongs to a specified *filter pool* and *profile*. The Filter Strings page tells you that this filter points to the home folder "." and that displayed file entries are filtered by name, but the file name filter, in fact, is "*" or everything. Note that none of this information is editable in the default filters that RSE creates.

You can create your own filters to simplify the management of remote resources. You might, for example, want a filter that lists just C/C++ source files, those with a .c, .h, or .cpp extension. Pick a folder on your remote system that has some source files in it, or in subfolders. In my case, it's drivers/ that has some examples for my Linux device driver class.

Right-click the folder name and select **New –> Filter** . . . This brings up the dialog in Figure 6.8. Select Subset by file types, and click **Select** . . . This brings up a list of file

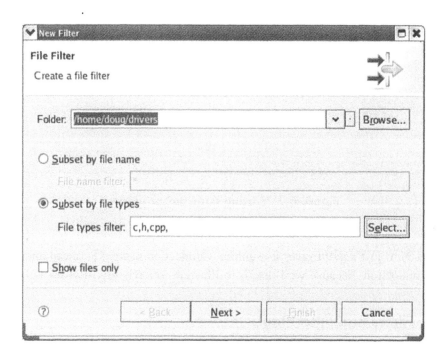

Figure 6.8: Creating a filter.

extensions, but sadly, .c and .h aren't listed. In Other Extensions enter "c,h,cpp" and click **OK**.

Click **Next** and give the filter a name, "My C Files," for example. On this screen you also have a choice of making this filter private to this connection or making it part of a filter pool, which makes it visible to other connections. Uncheck the **Only create filter in this connection** box to make it part of the default filter pool. Clicking **Next** brings up some "tips" about filters and filter pools. Click **Finish**.

The new filter shows up below Root under Sftp Files in your remote connection. For filters you create, you can modify the Filter Strings from the Properties page and add new filter strings.

Over time, you may end up with such a large number of filters that the system becomes hard to navigate. The solution is to aggregate related filters into filter pools. But in order to work with filter pools, we have to make them visible. Select **Window –> Preferences –> Remote Systems** and check the **Show filter pools in Remote Systems view** box. While you're there have a look at some of the other preferences related to remote systems. Click **Apply** and then **OK**.[1]

Back in the Remote Systems view, the filters previously listed under Sftp Files have been replaced by a filter pool entry called `<profile_name>:ssh.files`, where `<profile_name>` is the name of the default RSE *profile* that defaults to the name of your local host system. Expand the filter pool entry to reveal the same set of filters we previously had. Right-click the pool entry and select **Properties**. Here we see that this is not a filter pool but is, in fact, a *reference* to a filter pool.

Filter pools are associated with profiles and then referenced by connections. To create a new filter pool, right-click on Sftp Files and select **New –> Filter Pool** ... Give it a name, "My New Pool," perhaps. At this point there should be only one profile to which it can be attached named for your host computer. Click **Finish** and the new pool shows up under Sftp Files. You can now create new filters to add to this pool.

You might want to make your new filter pool visible to the second connection we created earlier. Right-click on the Sftp Files entry of your second connection and select **New –> Filter Pool Reference –> <your_host_name> –> My New Pool**. The new pool is now visible in your second connection.

---

[1] You can also manipulate this preference by clicking the **View Menu** icon in the Remote Systems view.

6.2.3   Profiles

Profiles are the "big picture," so to say, a way of tying everything together. RSE resources such as connections, filter pools, and other artifacts are owned by a profile. Figure 6.9 schematically illustrates how these elements tie together. Profiles are a useful tool for managing resources when you have a lot of connections.

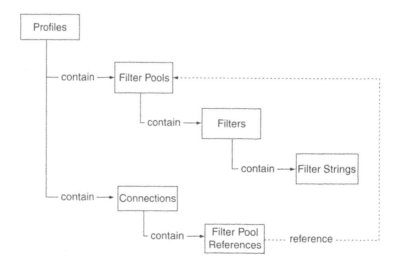

Figure 6.9: The role of profiles.

RSE creates an initial profile when it's started for the first time, usually named after the host name of the machine that creates the workspace. This profile is considered *private* and can't be deleted or made inactive. All of the objects we've created so far have gone into this profile.

Profiles are managed from the Team view, which should currently look something like Figure 6.10. You should see three connections: your local host, and two connections to a remote target.

Creating a new profile is trivial. Click the **New profile** icon in the Team view menu. All you have to do is give it a name and decide whether or not to make it active. Click **Finish** and the new profile immediately shows up in the Team view. The new profile has Connections and Filter pool entries, but of course, they're currently empty.

Now whenever you create a connection or a filter pool you have a choice of which profile to put it in.

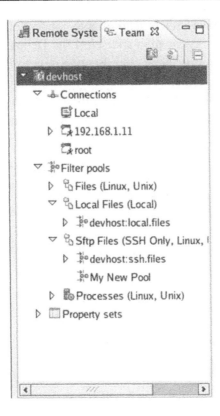

Figure 6.10: RSE Team view.

6.2.4   Debugging With a Remote Connection

With a remote connection established, we'll use the record_sort project to illustrate how to set up a remote debug configuration. Initially, you have to copy the application executable, `record_sort`, to a location on the remote target. My Home seems like a good place. The main reason for doing this is so you can change the permissions to make the file executable. Eclipse will, in fact, download the executable to the target when you click **Debug**, but if the file doesn't already exist, oddly enough it won't be created with execute permission. While you're at it, copy `datafile` to the target as well.

Now go back to the C/C++ perspective and open the Debug Configurations dialog for the record_sort project. You'll find a new configuration type called C/C++ Remote Application. Create a new configuration of this type.

In the Connection drop-down, select your remote connection. Then click **Search Project**... next to C/C++ Application and select the Debug binary just as you would for a local debug configuration. Now click **Browse** next to Remote Absolute File Path for C/C++. This brings up a dialog (Figure 6.11) that lets you specify where the application executable will be downloaded on the remote machine. Expand My Home and select the record_sort executable that you just downloaded.

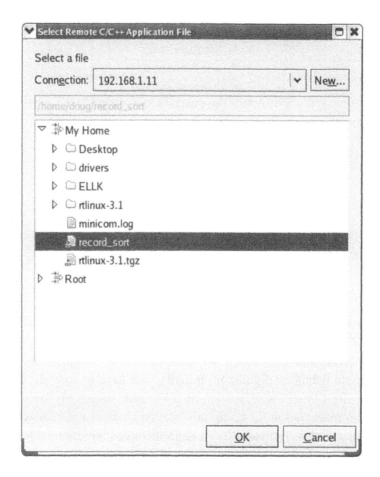

Figure 6.11: Select remote file location.

The Main tab of Debug Configurations now resembles Figure 6.12. Add "datafile" in the Arguments tab. Take a look at the Debugger tab and note that the only choice for Debugger is remote gdb/mi. Also note under Gdbserver Settings, that the default port number is 2345.

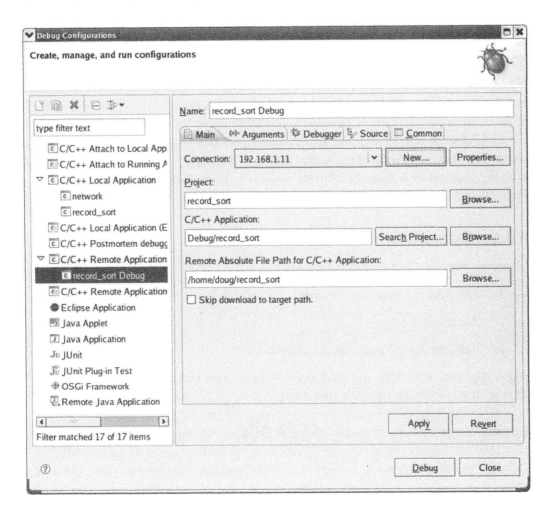

Figure 6.12: Remote debug configuration.

Click **Apply** and then **Debug**. Eclipse invokes make all, which may have nothing to do, downloads the executable to the target, starts gdbserver on the target, and switches to the Debug perspective. You're now debugging on the remote target without having to manually start gdbserver.

6.3   Native Application Builder (NAB)

The goal of the Native Application Builder (NAB) is to enable you to develop platform-independent graphical applications in an intuitive and graphical manner. Conceptually, this is similar to SWT and JFace, except that NAB is written in C++ and is oriented toward developing C++ applications.

NAB makes use of run-time libraries from WideStudio/MWT, an open source project based in Japan. The MWT stands for Multi-platform Widget Toolkit. WideStudio/MWT is described as an "Integrated Development Environment for desktop applications." Nevertheless, the intention is that applications developed with WideStudio should be able to run on a wide range of embedded platforms simply by recompiling and relinking with the appropriate library. In addition to C/C++, WideStudio supports Java, Perl, Ruby, Python, and Objective Caml (OCaml).

6.3.1   Getting and Installing NAB

There are actually three elements to NAB:

1. The NAB Eclipse plug-in itself

2. WideStudio

3. JDK, the Java 2 software development kit

These elements need to be installed in the reverse order that they're listed above. That is, JDK should be installed first. Go to http://www.java.sun.com/javase/downloads/index.jsp and click **Download** for **JDK 6 Update 7** (or whatever is the latest update at the time). On the next page select Linux as your platform (unless you're running Eclipse under Windows), agree to the Java SE Development Kit 6 License Agreement, and click **Continue**. You are then presented with a choice of downloading a self-extracting binary or a self-extracting RPM. I chose the binary.

Move the resulting `.bin` file to the directory where you want to install it. I chose `/usr/local/`. Then execute it. The binary code license is displayed, and you are prompted to agree to its terms. The JDK files are installed in a directory called `jdk1.6.0_<version>` in the current directory.

The DSDP-NAB project downloads page has a link to the WideStudio MWT libraries package, which happens to be hosted at http://www.sourceforge.net/.

Download the file, and then follow these instructions to install it:

1. Copy the `*.tar.gz` file to the directory where you want to build it.

2. `cd` to that directory.

3. `cd ws-v3.97-12/src`

4. `./configure`

5. `make runtime`

6. `export JAVA_HOME=<JDK_install_directory>`

7. `make mwt_java`

8. `su`

9. `make install`

10. Add `/usr/local/lib` to the `LD_LIBRARY_PATH` environment variable.

11. Add `/usr/local/ws/bin`[2] to your `PATH`

Step 6 is necessary if, like me, you have an older version of the JDK lurking in the default location.

The final step is installing NAB, which doesn't show up in any of the current software update sites, so you would expect to install it "manually." This is what the DSDP-NAB download page describes. Unfortunately, it doesn't work. There are apparently some undocumented dependencies in the NAB plug-in.

A plea for help on the DSDP-NAB newsgroup finally yielded a response that included the link to an NAB update site:

http://download.eclipse.org/dsdp/nab/updates/

Add this to your list of update sites and expand it. You'll find three entries:

- NAB/MWT for Linux GTK

- NAB/MWT for Win32

- Uncategorized, which expands to:

  o NAB/MWT Plug-in

---

[2] Regardless of where you chose to build the package, the MWT runtimes get installed in `/usr/local` by default.

Select the third item and one of the first two depending on your host platform.

Note: It seems that many, if not most, of the mature Eclipse projects have established update sites, although it's not always apparent. It may take some serious digging around and asking questions on the relevant newsgroup. As a last resort, see if the following link exists:

http://download.eclipse.org/<project_name>/<sub_project>/updates/

where <project_name> is the name of the project you're trying to install and <sub_project> is an optional sub-project name.

6.3.2 An NAB Project

After adding the NAB components, you'll find a new perspective called NAB/MWT that initially looks something like Figure 6.13. The NAB/MWT perspective introduces several new views. You may want to expand Eclipse to full screen in order to see more detail.

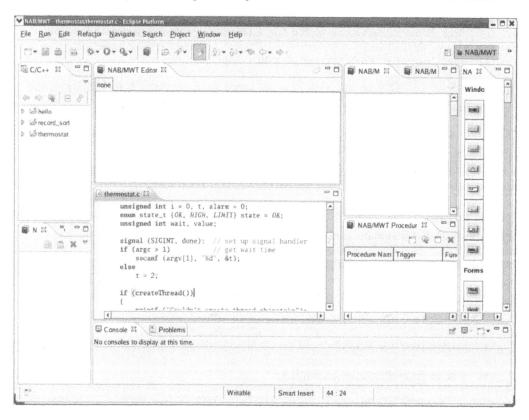

Figure 6.13: Initial NAB/MWT perspective.

Proceeding from left to right, the views are:

- **C/C++ Projects**: This is virtually identical to the Project Explorer view in the C/C++ perspective. Underneath it are . . .

- **NAB/MWT Tree**: Shows the current project's application windows and their elements, or *instances*, in tree form.

- **NAB/MWT List**: For each instance in the Tree view, this view shows the child instances. Moving to the right . . .

- **NAB/MWT Editor**: Enables you to visually lay out screen images, known as *application windows*. Multiple windows are identified by tabs across the top.

- **NAB/MWT Properties**: Displays and edits the properties of an instance selected in the Editor, Tree, or List.

- **NAB/MWT Attributes**: Displays and edits the attributes of a selected instance. The only attribute for a variable is Global. A window instance has a Type that identifies it as a Normal window, a Class definition, or an object store that isolates window configuration information in a file.

- **NAB/MWT Procedures**: Associates *event procedures* written in a procedural language with instances. A procedure has an arbitrary Procedure name, a Function name by which it is called, and a Trigger that specifies the conditions for calling it.

- **NAB/MWT Object Box**: Displays icons for all object types and classes implemented by WideStudio/MWT. These are added to an application window by simply dragging them into the Editor.

So let's create our first NAB project. Select **File –> New –> Project** and you'll find a new wizard category called NAB. Expand that and select **NAB/MWT Project** (Figure 6.14). Next, you're asked to select between C++ and Java as the Programming language. Leave it as C++. Then you give the project a name. Call it "HelloNAB" and click **Finish**.

The new project shows up in the C/C++ Projects view, and already has a number of files including a template .ccp file. In fact, Eclipse went ahead and built the project, although there isn't anything useful in it yet. Have a look at `HelloNAB.cpp` just to get a feel for what's there. Remember that the objective of NAB is to generate programs that make use of the WideStudio/MWT libraries.

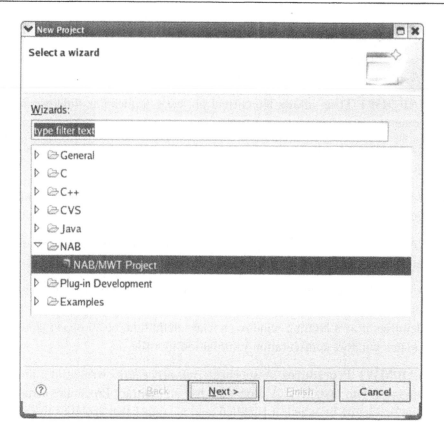

Figure 6.14: Select NAB project wizard.

Creating an Application Window

Select **File** –> **New** –> **Other** ... The wizard dialog now has another entry under NAB called NAB/MWT Window. Select that and click **Next**. The new window gets a name, a class, and a type. For now, leave the defaults and click **Finish**. The new window shows up in the NAB/MWT Editor and the Properties view shows that it is 400 × 400 pixels. Add a Title string property. Call it "NAB Application."

Now we'll add a button to the window. Find the Commands section in the Object Box view and select the first entry, WSCvbn (Button class). Select that and click somewhere near the upper left corner of the `newwin001` tab of the Editor. A new button instance is created. Grab a corner of the button and expand it. Note that the Properties view reflects the current location and size of the button, among other things. Edit the properties (in the Properties view) as follows:

X:	10
Y:	10
Width:	200
Height:	30
Label string:	Hello

The result should look like Figure 6.15.

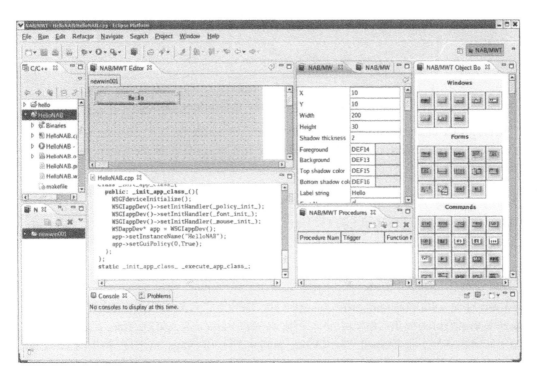

Figure 6.15: Application window in NAB/MWT perspective.

Creating an Event Procedure

In order for our new button to actually do something, we have to attach it to an event procedure. With the Hello button selected in the Editor, click the **Add procedure** icon in the Procedures view to bring up the dialog in Figure 6.16. Name the function "hello_button" and select ACTIVATE as the trigger.

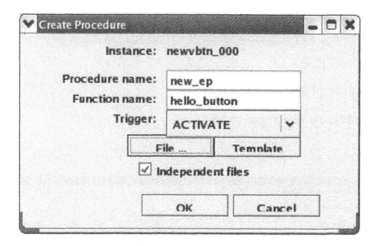

Figure 6.16: NAB Create Procedure dialog.

Clicking **OK** creates a new .cpp file with a template for the `hello_button()` function. In the source code Editor, enter the following line in the `hello_button()` function:

```
object->setProperty (WSNlabelString, "Hello from NAB");
```

When you save the revised `hello_button.cpp` file, the project is rebuilt. Now we have a program that does something.

Running the Program

As we have with previous projects, open the Debug Configurations dialog and create a new configuration for this project. The default values should be fine. Click **Apply** if it's active and click **Debug**. The Debug perspective displays an error in an Editor tab saying that no source is available for "main()". That's because `main()` is supplied by the WideStudio/MWT library. That's OK, because we're not interested in debugging `main()`.

Click the **Resume** icon. The application window with the Hello button shows up (Figure 6.17). When you click the button, the label changes to the string you set in the `hello_button()` function.

Figure 6.17: The running application window.

Granted this is not a very exciting application, but it does serve to illustrate the fundamental capabilities of NAB. You can easily build and test the GUI elements of an embedded application on your host and then move it over to the target. In a sense, this is a more elaborate version of the simple simulation we did in the previous chapter.

NAB Project Properties

While building a native application is an instructive exercise, the real power of WideStudio/MWT and the NAB plug-in is the ability to rebuild the project for an embedded target environment. The target environment is configured and described in the project Properties dialog.

When you select Properties for an NAB project, you'll find three additional entries in the navigation panel:

- NAB/MWT Platform SDK Environment

- NAB/MWT Project Class Library Settings

- NAB/MWT Project Settings

For now, we'll take a look at the Project Settings dialog shown in Figure 6.18. There are three tabs. Target Settings has three items. Encoding lets you specify a character encoding if one is not explicitly specified in the program. Application type is one of Normal Application, Class Library, Netscape Plugin, or Console Application.

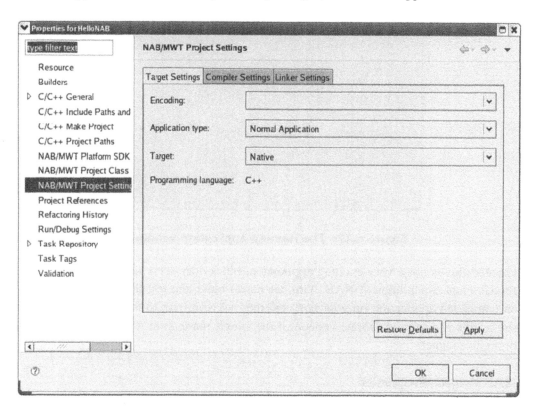

Figure 6.18: NAB/MWT Project Settings dialog.

The Target drop-down brings up a very extensive list of all the platforms supported by WideStudio/MWT (Figure 6.19). Of course, selecting anything other than Native

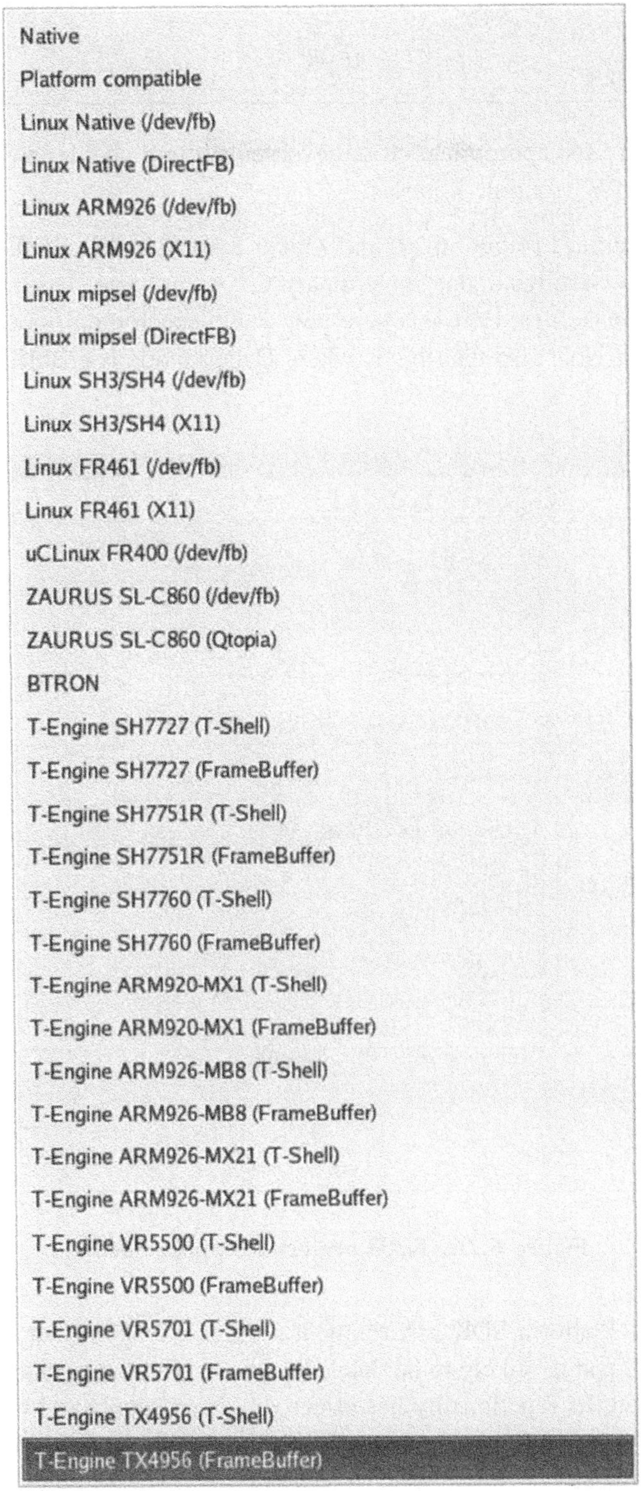

```
Native

Platform compatible

Linux Native (/dev/fb)

Linux Native (DirectFB)

Linux ARM926 (/dev/fb)

Linux ARM926 (X11)

Linux mipsel (/dev/fb)

Linux mipsel (DirectFB)

Linux SH3/SH4 (/dev/fb)

Linux SH3/SH4 (X11)

Linux FR461 (/dev/fb)

Linux FR461 (X11)

uCLinux FR400 (/dev/fb)

ZAURUS SL-C860 (/dev/fb)

ZAURUS SL-C860 (Qtopia)

BTRON

T-Engine SH7727 (T-Shell)

T-Engine SH7727 (FrameBuffer)

T-Engine SH7751R (T-Shell)

T-Engine SH7751R (FrameBuffer)

T-Engine SH7760 (T-Shell)

T-Engine SH7760 (FrameBuffer)

T-Engine ARM920-MX1 (T-Shell)

T-Engine ARM920-MX1 (FrameBuffer)

T-Engine ARM926-MB8 (T-Shell)

T-Engine ARM926-MB8 (FrameBuffer)

T-Engine ARM926-MX21 (T-Shell)

T-Engine ARM926-MX21 (FrameBuffer)

T-Engine VR5500 (T-Shell)

T-Engine VR5500 (FrameBuffer)

T-Engine VR5701 (T-Shell)

T-Engine VR5701 (FrameBuffer)

T-Engine TX4956 (T-Shell)

T-Engine TX4956 (FrameBuffer)
```

Figure 6.19: NAB/MWT

assumes you have the appropriate cross-development tool chain, and have built the corresponding MWT runtime libraries.

The Compiler Settings (Figure 6.20) and Linker Settings tabs take the place of the **C/C++ Build –> Settings** dialog for ordinary C/C++ projects. This is because NAB creates its own makefiles. Here is where you would specify a cross compiler for your embedded target. Note also that this is where Debug mode is enabled.

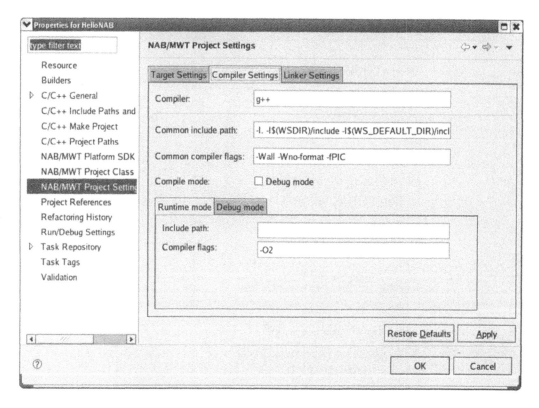

Figure 6.20: NAB project Compiler Settings.

The NAB/MWT Platform SDK Environment settings are an artifact of the original WideStudio IDE and are likely to be deleted in the next major update, according to the NAB project lead. Its functionality has effectively been supplanted by the Project Settings. Incidentally, T-Engine is a popular embedded platform in Japan that grew out of the TRON project.

6.4   Other DSDP Subprojects

Although they're still in the incubation stage, it's worth at least describing the remaining subprojects under the Device Software Development Platform. If you're developing mobile devices, you might want to look into them.

6.4.1   Mobile Tools for Java (MTJ)

According to the project's webpage, the goal of the Mobile Tools for Java (MTJ) project is to extend existing Eclipse frameworks to support mobile device Java application development. The intention is to develop frameworks that can be extended by tool vendors and tools that can be used by third-party mobile Java application developers.

At the beginning of 2008, the project was "rebooted." It went from version 0.7 to 0.1. The goal now seems to be to extend the functionality of EclipseME version 1.7.9.

6.4.2   Tools for Mobile Linux (TmL)

The Tools for mobile Linux (TmL) project intends to "address the gap where existing Eclipse projects do not entirely satisfy the needs of developers of applications for mobile devices," according to the project's webpage. The initial scope focuses on building a device emulator framework supported by a VNC Viewer for graphic display visualization and a simulated end-to-end environment to test enterprise applications.

Summary

In this chapter we've looked at a couple of features of the Device Software Development Platform (DSDP), an Eclipse project specifically focused on issues of embedded software development. We started out by looking at how Eclipse handles software updates and installing extensions. Using update sites makes the process of extending Eclipse relatively painless and transparent because the update mechanism takes care of resolving any dependencies in the software you want to install. You can also instruct Eclipse to automatically check for new updates to installed software on a regular basis.

DSDP itself is divided into several subprojects. The Remote System Explorer (RSE) is a collection of tools that allows you to work with resources such as files and folders on remote systems. You can copy files to and from the remote system with a simple drag-and-drop paradigm, edit files directly on the remote system, and run and debug applications remotely.

The Native Application Builder (NAB) enables you to develop platform-independent graphical applications in an intuitive and graphical manner. NAB is an Eclipse wrapper for WideStudio/MWT, an open source project focused on developing platform-independent graphical applications for embedded devices. You can create and test your application on your host environment initially, and then rebuild it for the appropriate target environment.

There are two other subprojects under DSDP that focus on aspects of mobile device development. These are still in a very early "incubation" phase.

Up to this point we've talked about how to use the features that are already available in Eclipse. Suppose you want to add some new functionality to Eclipse. The next chapter looks at the process of developing Eclipse plug-ins.

CHAPTER 7

Plug-In Development Environment (PDE)

As has been mentioned before, Eclipse is built on the notion of plug-ins that define its functionality beyond the base platform. So in order to extend Eclipse's functionality, you need to create one or more additional plug-ins. As an embedded C developer, you may never need to extend Eclipse, but if you do, this chapter will serve as a starting point.

Perhaps not surprisingly, Eclipse provides a powerful, easy-to-use tool, the Plug-in Development Environment (PDE), to assist in creating new plug-ins. PDE hides many of the excruciating details of plug-in development.

7.1 Installing the PDE

PDE itself is of course a collection of plug-ins that must be downloaded and installed into Eclipse just as we did with the DSDP in the last chapter. PDE depends on the Eclipse Java Development Toolkit, JDT. That, too, needs to be downloaded and installed. The JDT in turn depends upon the Java Development Kit (JDK) from Sun Microsystems.

The update site for the Plug-in Development Environment is part of the Ganymede update bundle under Java Development. Just select Eclipse Plug-in Development Environment and click **Install**... JDT and any other plug-ins that PDE requires will be installed.

There does appear to be a "gotcha" that wasn't in Eclipse 3.3. PDE does not recognize that it needs the Eclipse SDK and Eclipse Platform SDK in order to function, so you have to explicitly install those from the Eclipse Project Update Site.

The final step in installing PDE is to install the Java Development Kit from Sun Microsystems. Go to http://java.sun.com/javase/downloads/index.jsp and **Download ->** **JDK 6 Update 7**. On the next page select Linux as your platform (unless you're running Eclipse under Windows), agree to the Java SE Development Kit 6 License Agreement, and click **Continue**. You are then presented with a choice of downloading a self-extracting binary or a self-extracting RPM. I chose the binary.

7.2 So What Is a Plug-In?

It might help to think of Eclipse as the software equivalent of a USB hub. Plug a device into a USB hub and the system automatically figures out what it is and how to drive it. Likewise, when a plug-in is installed into Eclipse, the system determines what the plug-in is capable of doing and what it depends on, so that things get loaded in the proper order.

The core of a plug-in is Java code, so in order to develop truly useful plug-ins, you'll need to know Java. This book is not the place to learn it. The Resources section at the end of the chapter can point you to Java resources.

With the release of version 3.0, Eclipse adopted the OSGi[1] framework for modular, dynamic, Java components. The OSGi framework defines a dynamic component model that allows applications or components, in the form of "bundles," to be remotely installed, started, stopped, updated, and uninstalled without requiring a system reboot. An Eclipse plug-in is the equivalent of an OSGi bundle.

7.2.1 Extensions and Extension Points

Fundamentally, extensions and extension points are how plug-ins communicate with the Eclipse platform and with each other. Conceptually, you can think of an extension point as a socket and an extension as a plug that mates with the socket.

Extension points define places where the Eclipse platform can be extended by plugging in additional functionality in the form of extensions. The current Eclipse platform exposes around 200 extension points in the following categories:

- **Platform Runtime**: Controls the global behavior of Eclipse itself. If you wanted to write an application based on the Eclipse platform, this is the place to start.

---

[1] OSGi originally stood for Open Services Gateway initiative. At some point that term was dropped, and now the organization and its specifications are simply referred to as OSGi.

- **Workspace**: Contains extension points for managing projects, such as resource builders (build tools), markers, project life-cycle behavior, and team behavior.

- **Platform Text**: Facilities for extending editors.

- **Workbench**: Manages the user interface. This is the largest class of extension points and includes things like support for new views and editors, key bindings, drag-and-drop operations, and additional panels in the Preferences dialog.

- **Team**: Extension points to manage sharing of project resources, such as folders and files, among team members.

- **Debug**: Controls application launching as well as debugging. Includes extension points for both behavior and user interface functionality.

- **Console**: A set of three extension points for managing console views.

- **User Assistance** (in other words, Help): Facilities to extend the Eclipse Help engine. You can add new HTML help pages and cheat sheets, extend the table of contents, and access the search engine.

- **Language Toolkit**: Extension points that support refactoring.

- **Other**: Provides access to the underlying Ant infrastructure. Facilities include read-only content viewers and specialized search pages.

The Platform Plug-in Developer Guide in the Eclipse online documentation provides a thorough description of all the extension points and this is, in fact, the best place to get familiar with what the various extension points represent.

Extensions in turn provide the additional functionality that plugs into the extension points. This is what plug-ins generally do. They extend Eclipse functionality by "plugging in" to extension points.

7.2.2 MANIFEST.MF and plugin.xml

In Eclipse terminology, plug-ins are defined by two files, MANIFEST.MF and plugin.xml. The MANIFEST.MF file identifies the plug-in by name, version, and so on. The plugin.xml file defines the extensions and extension points that this plug-in provides. This is how the plug-in gets linked into the Eclipse platform environment.

Ultimately, `MANIFEST.MF` and `plugin.xml` get packaged along with the Java code in a Java archive (.jar) file.

7.2.3 Naming Conventions

You've probably noticed by now that all of the Eclipse plug-ins have rather long names of the form `org.eclipse.<feature>.<element of feature>_<number that looks like a version><maybe another big long number>.jar`. This is a naming convention that assures that all Java packages, or in this case Eclipse plug-ins, are unique. Eclipse requires that all plug-ins have a name that is unique throughout the world.

The name begins with a top-level Internet domain, such as .com, .org, .edu, etc., or one of the two-letter codes that identify a country outside the United States. The next element is a domain name representing the organization that wrote and/or maintains the package. Subsequent components of the name are specified by the organization and may include elements such as division, department, project, and so on.

For Eclipse the next component of the name is a component such as `core`, `ui`, or `debug`, or a subproject such as `cdt`, `jtk`, or `pde`. This is usually followed by the element of the component or subproject. Then comes the underscore followed by a version number. This is followed by another underscore, a "v", and the date the package was built.

The convention further suggests that the Eclipse project name should match the plug-in name up to the first underscore.

7.3 Our First Plug-In

As a way to get started with writing plug-ins, we'll create a simple workbench view that just lists all views available in the workbench at run-time. You might think of it as a "super view" so that's what we'll call it: SuperView. While it doesn't really do a lot, it does illustrate most of the basic principles involved in using PDE.

7.3.1 Creating a Plug-In Project

Like every other project, building a plug-in starts by creating a project. Select **File –> New Project**. Find and expand the **Plug-in Development** category and select **Plug-in Project** as shown in Figure 7.1.

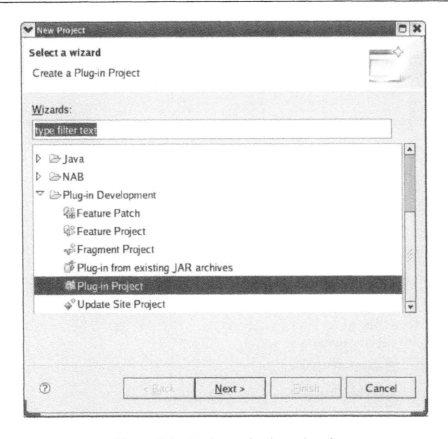

Figure 7.1: Project selection wizard.

This brings up the dialog shown in Figure 7.2. Again, like every other project, we have to give it a name. In accordance with the Java naming standard, let's call it `com.eclipsebook.superview`. There is also an option to choose a target platform, which simply means whether we're targeting a version of Eclipse, or an OSGi framework like Equinox. In this case, we'll stick with the 3.4 version of Eclipse to keep things simple.

The next screen (Figure 7.3), called **Plug-in Content**, basically identifies the plug-in in the form of properties, and says something about its content. The properties are structured to make the plug-in compliant with Eclipse. All the default values are suitable.

The Plug-in ID represents a unique identifier for your plug-in. No other plug-in can share the same identifier. Note that the default value for the ID is the same as the name

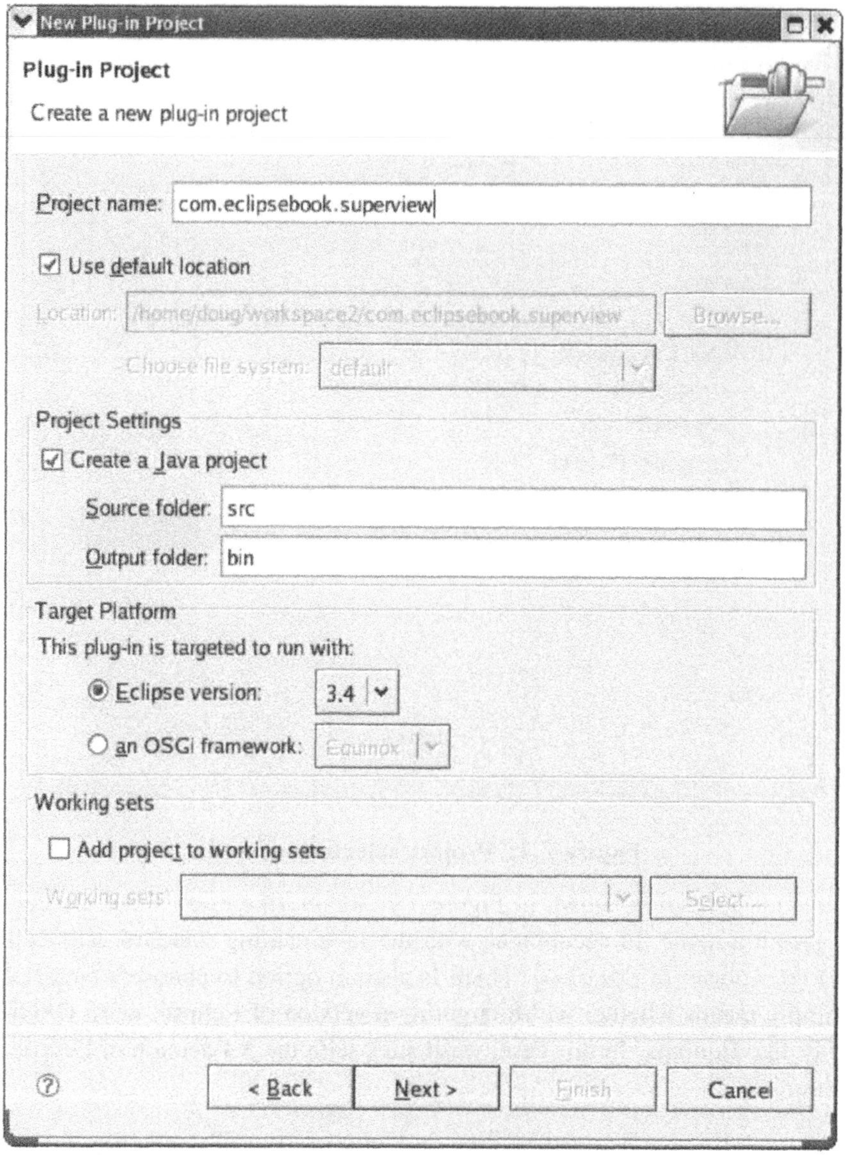

Figure 7.2: Plug-in new project dialog.

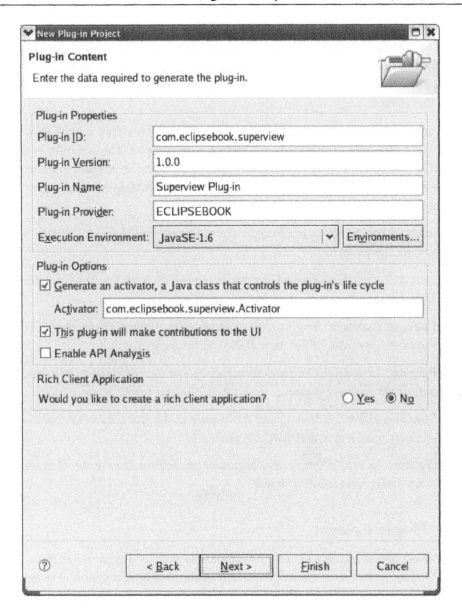

Figure 7.3: Plug-in Content dialog.

we used in the previous screen. The Plug-in Version consists of four segments (three integers and a string) respectively named *major.minor.service.qualifier* The Plug-in Name is simply a human-readable name. Likewise, the Plug-in Provider is a human-readable string identifying the author of the plug-in, which defaults to the organization field of the name.

There's an option to generate a *plug-in activator*, which is simply a class that controls the life cycle of a plug-in; basically a start and stop method. The activator is usually responsible for setting up things and for properly disposing of resources when the plug-in isn't needed anymore. We'll also say here that this plug-in makes contributions to the UI.

Click **Finish**. If you're not in the Plug-in Development environment, Eclipse will prompt you to switch to it.

PDE has created a template project for our new plug-in. Views initially visible in the Plug-in Development Perspective include:

- **Package Explorer**: The Java equivalent of the Project Explorer in CDT. It shows the contents of our new Hello World project.

- **Plug-ins**: Lists all of the plug-ins presently installed in Eclipse. You can select a plug-in and open it in the. . .

- **Manifest Editor**: A form-based multi-page editor that aids in the creation of the `plugin.xml` and `MANIFEST.MF` files.

- **Outline**: As in CDT, this view organizes the information in the Manifest Editor in an easily browsable structure.

7.3.2   Plug-In Content

The Manifest Editor initially shows an overview as seen in Figure 7.4. This essentially duplicates the information entered in the Plug-in Content dialog and also includes links to several other tabs in the Manifest Editor and related functionality. Our plug-in will create a new extension that contributes to the Eclipse toolbar and also adds an item to the main menu.

The **Dependencies** tab specifies what plug-ins and packages this plug-in depends on. Initially the **Required Plug-ins** list has two entries: `org.eclipse.ui` and `org.eclipse.core.runtime`. Just about everything is dependent on these two.

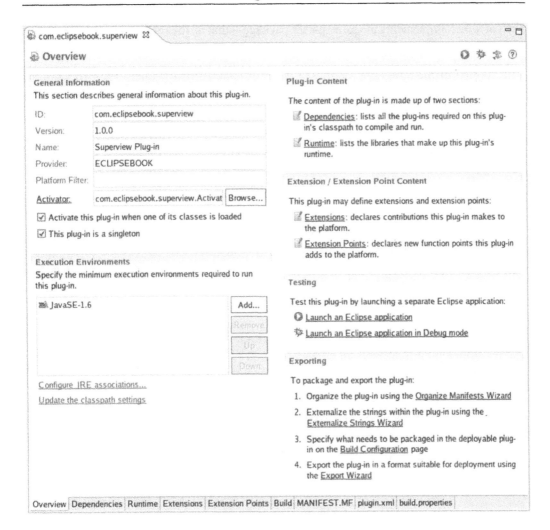

Figure 7.4: Manifest Editor overview tab.

The **Imported Packages** list lets you specify packages that your plug-in depends on without identifying the plug-in that contains the package.

Packages are a useful way to specify implementation-independent behavior. Suppose, for example, that your plug-in is dependent on an XML parser. We can implement that parser as a package called `com.company.xml.parser`. You could then create two plug-ins, say, `com.company.xml.parser.mobile` and `com.company.xml.parser.desktop`, each of which implements the same parser package—that is, the same extension points, but for a different environment.

Click the **Runtime** tab in the Manifest Editor, or click **Runtime** under **Plug-in Content** in the Overview form. This lets us specify which packages contained in the plug-in will be made visible to other plug-ins. Any packages not explicitly exported will not be visible externally. Click **Add**... to see that our plug-in has one package available for export. We don't really need to export the package, but it doesn't hurt, so go ahead and select it, and click **OK**. Now click on the package in the Exported Packages list. The **Properties**... button lets you set a version for that package.

Normally, exported packages are visible to all "downstream" plug-ins. Nevertheless, you can declare packages to be *internal*, meaning they're not necessarily intended for use by downstream plug-ins, but are still visible. Such internal packages can be hidden when Eclipse is started in "strict" mode by selecting the **hidden** option under package visibility. Internal packages can also have "friend" plug-ins that have access to the package even in strict mode.

The **Classpath** section lists all locations the runtime will search when loading classes from a plug-in. This is an artifact of earlier Eclipse releases and should normally be left blank. Starting with Eclipse 3.1, plug-ins are packaged as Java archives (.jar) and don't require an extended classpath.

Click the **Extensions** tab or click **Extensions** in the **Extension / Extension Point Content** section of the **Overview** tab. This is where the real functionality of our plug-in is defined. Clicking **Add**... brings up a list of extension points defined in the plug-ins listed in the **Dependencies** tab. Here we select the extension points for which our plug-in will provide extensions. Find and select `org.eclipse.ui.views`. A description of the extension point appears (Figure 7.5). Many extension points also offer one or more templates to provide additional help in structuring the plug-in's Java code. In this case, we won't use the template. Click **Finish**.

Back in the **Extensions** list, right-click on the new entry and select **New** –> **category** (Figure 7.6). Leave the **id** as is, and change the **name** to EclipseBook.

Right-click the extension point entry again and select **New** –> **view** (Figure 7.7). Change the name to SuperView and add `com.eclipsebook.superview.category1` to the category to match the id in the category entry. Click the **class** link, enter SuperView as the name, and click **Finish**.

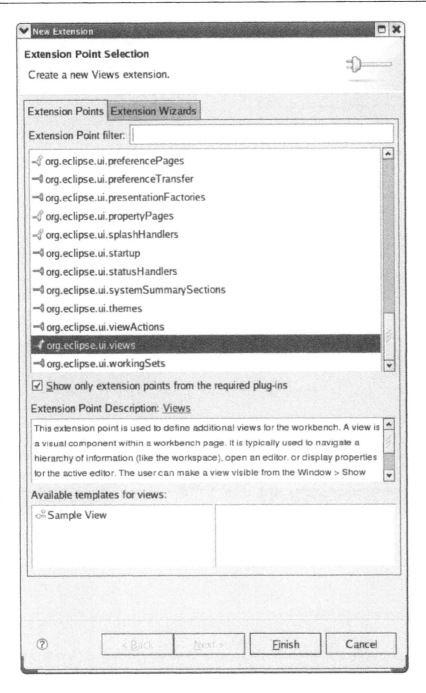

Figure 7.5: Selecting extension points.

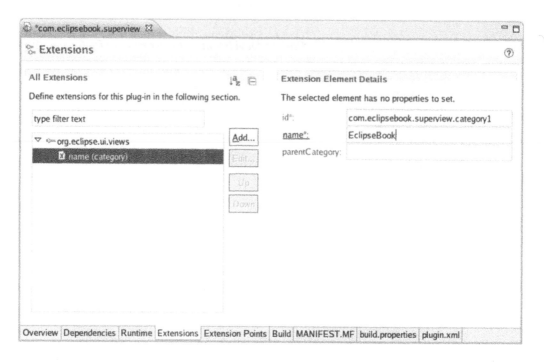

Figure 7.6: Adding a category.

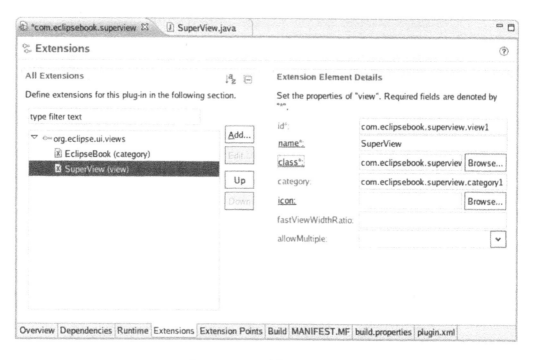

Figure 7.7: Adding a view.

The PDE creates a new class implementing the required interface and opens it in the Java editor. We'll come back to that in a moment, but for now let's finish up our tour of the Manifest Editor.

The final product of the Manifest Editor is the two files that constitute the plug-in's manifest, `MANIFEST.MF` and `plugin.xml`. The content of these files is available in tabs of the same names. Have a look. `MANIFEST.MF` basically describes the plug-in with the information from the Overview and Dependencies forms. `plugin.xml` defines the extensions and extension points that this plug-in provides. In this case, we haven't defined any extension points.

You can, of course, edit `MANIFEST.MF` and `plugin.xml` manually, but why? The beauty of PDE is that you don't need to understand XML or the syntax of `MANIFEST.MF`, in order to create a plug-in. Some developers even suggest that it is bad practice to manually edit an XML file as this can lead to hard-to-diagnose configuration problems.

Now go back to the Editor tab displaying `SuperView.java`. Not much here. Just templates for the class constructor and two other functions that are the minimum necessary to extend the view. To get something that works, import the file `SuperView.java` from `EclipseSamples/plug-in/` to `com.eclipsebook.superview` in the Package Explorer. Eclipse asks if you want to overwrite the existing file. Yes, you do.

7.3.3 Running and Debugging a Plug-In

Eclipse and PDE have a notion of *self-hosting*, which means that we can launch a new instance of Eclipse with plug-ins we're currently working on, without having to export or deploy any plug-ins. One way to start the new instance is to go back to the Overview form in the Manifest Editor and click on **Launch an Eclipse application** under **Testing**. You can also just click on the **Run** icon in the menu toolbar or, from the top-level project context menu in the Package Explorer view, select **Run As –>
Eclipse Application**.

After a little churning, a new instance of Eclipse appears in the default C/C++ perspective. Interestingly, it doesn't ask for a workspace location. Instead, it creates a workspace in your home directory, called runtime-EclipseApplication.

In the main menu select **Window –> Show View –> Other**. Expand the **EclipseBook** entry and select **SuperView**. The new view shows up in the bottom window.

Now that you know what the plug-in does, go back into the Java code and try to figure out *how* it does it.

Debugging is virtually identical to what we did with CDT. The one "gotcha" here is that there's no "Stop on main." This means you must have a breakpoint set before you launch the debug run. Open `SuperView.java` if it's not already open. Now switch to the Debug perspective. This is essentially the same Debug perspective we used for C programming, but now we're using a Java debugger instead of gdb.

Scroll down to line 24 in `SuperView.java` that starts with `viewer.setContentProvider...` and set a breakpoint just as we did before. Now from the Manifest Editor Overview form, click **Launch an Eclipse application in Debug mode**. The new Eclipse instance starts as before. Select the **SuperView** view as we did above and the program hits the breakpoint (Figure 7.8).

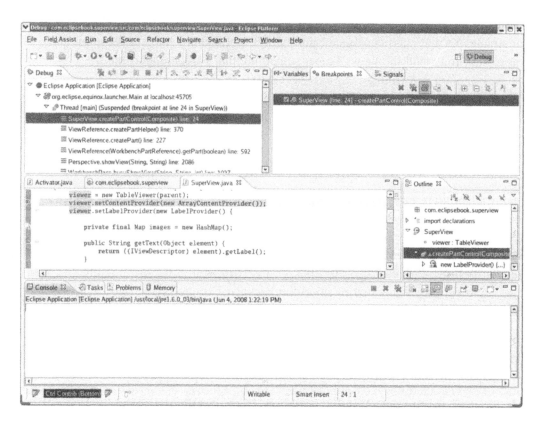

Figure 7.8: Java Debug perspective.

The Debug view shows the call stack with the method containing the breakpoint highlighted. If you select any other frame in the call stack, a window appears in the editor saying that the source is not available. Nevertheless, the local variables for that frame are displayed in the Variables view and the Outline view shows an outline of the corresponding method.

Note, incidentally, that the debug instance of Eclipse is blank and waiting for something to happen. You can step through the `createPartControl()` method to see what happens, but in fact nothing actually changes on the workbench until after `setFocus()` is called.

Without going into a lot of detail, let's take a quick look at what `createPartControl()` does. First it creates a table-like visual component, a "table viewer," to display your view's contents. It then configures the viewer with a content provider and a label provider. The former allows the viewer to navigate your model, or more specifically, to extract your model's structural elements suitable for inclusion in a table. The latter allows it to convert your model's elements into table cells with textual labels and optional images.

Then there's a sorter to display the entries in alphabetical order. Finally, you provide an input source that is an array of descriptors of all views available in the workbench at run-time. Note that we're providing an implementation of `setLabelProvider()` here, and we're overriding the `getText()` and `getImage()` methods.

7.4   Building and Exporting a Plug-In

Our plug-in is fully debugged and it's time to share it with the rest of the world. This requires packaging all of its components in a Java archive (.jar). This, too, is a fairly painless process owing to wizards available to help.

There are three steps to going from a working plug-in to its corresponding .jar file:

- Clean up and organize the manifests.
- Configure the build content.
- Build and export the plug-in.

7.4.1   Clean Up and Organize the Manifests

PDE provides an **Organize Manifests** wizard to help ensure that the information in your `MANIFEST.MF` file is complete and up to date. Among other things, it gives you the opportunity to override and extend options selected in the Manifest Editor.

The **Organize Manifests** wizard (Figure 7.9) is accessible from the Overview form in the **Exporting** section. The various options, which tend to be rather verbose, are:

- **Ensure that all packages appear in the MANIFEST.MF**: Adds Export-Package declarations for any package in the project that isn't already exported.

- **Mark as internal all packages that match the following filter**: Can change the visibility of a package based on its name. The Package filter field is a regular expression that identifies packages to be marked as internal. Internal packages are not available for use by other plug-ins.

- **Remove unresolved packages**: Removes Export-Package entries that can't be resolved, probably because they simply don't exist in the project.

- **Calculate 'uses' directive for public packages**: Does the complex computations for the "uses" directive for packages that other plug-ins have access to. This requires code introspection and can be a lengthy process.

- **Handle unresolved dependencies by**: Offers the option of removing unresolved dependencies or declaring them as optional. The most common reason a dependency might be unresolved is that an optional plug-in is missing from the configuration.

- **Remove unused dependencies**: Analyzes the project code for references to unused dependencies, which could result in unnecessary plug-ins being installed at runtime. This can be a time consuming process.

- **Add required dependencies**: Sort of the complement of the previous option, it looks for dependencies that are not currently included in the manifest. This only finds dependencies for plug-ins listed in the **Automated Management of Dependencies** section of the Dependencies form of the Manifest Editor. This can be a time consuming operation.

- **Remove unnecessary lazy activation headers**: Removes so-called "lazy activation headers" if a Bundle-Activator is not defined. The lazy activation header isn't necessary if a plug-in has nothing to contribute when it is started.

- **Delete unnecessary plug-in manifest files**: Removes `plugin.xml` files if a plug-in doesn't contribute any extensions or extension points.

- **Prefix icon paths in plug-in extensions with a nl segment**: Modifies icon paths to allow fragments to contribute unique icons for different locales.

- **Remove unused keys from the plug-in's properties file**: Finds and removes unused keys in the properties file.

- In most cases the default values are fine.

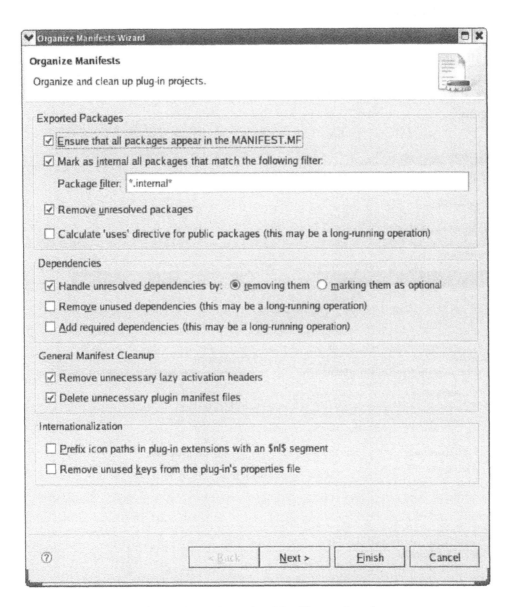

Figure 7.9: Organize Manifests wizard.

7.4.2   Configure Build Content

The file `build.properties` defines what is included in the final plug-in. Its contents are specified in the Build form of the Manifest Editor shown in Figure 7.10. Again, for now anyway, the defaults are fine.

The **Binary Build** is what you specify for building your plug-in. Oddly, the `bin` and `src` directories are not selected, yet they do show up in the actual text of the `build.properties` file. The **Source Build** is not commonly used or needed by general users. It's used if you need to ship source in separate plug-ins rather than in the binary plug-ins.

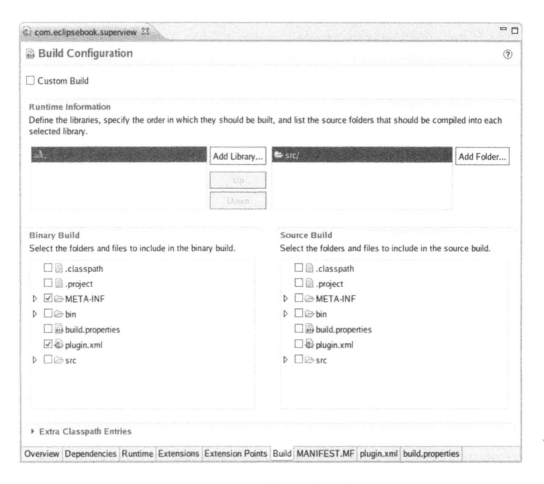

Figure 7.10: Build Configuration.

7.4.3 Build and Export

The final step in creating a deployable plug-in is to build the `.jar` file and place it somewhere in the file system. The **Export Wizard** is the last item under **Exporting** on the Manifest Editor's Overview form. Clicking that link brings up the dialog in Figure 7.11. The wizard lists the plug-ins that are available to deploy, in this case just our com.eclipsebook.superview project. Check it if it isn't already checked.

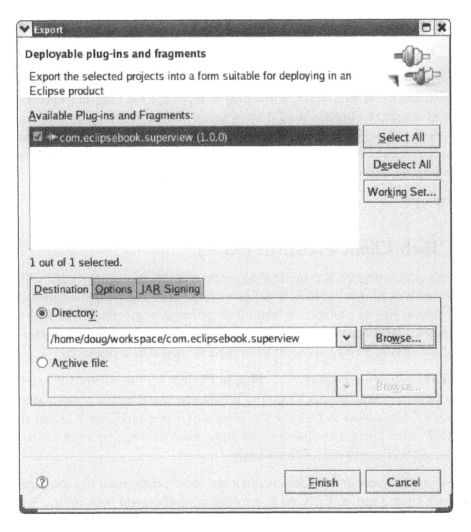

Figure 7.11: Plug-in export wizard.

Select a destination. I chose to store it back in the Superview project folder in the workspace. The wizard stores the `.jar` file in a subdirectory named `plugins/` under the directory you specify. If you're curious, take a look at the **Options and JAR Signing tabs**. Click **Finish** and the plug-in will be built.

The plug-in export wizard can also be accessed from **File –> Export**. Select **Plug-in Development –> Deployable plug-ins and fragments**.

That's it! You've created and deployed an Eclipse plug-in.

7.5   Exploring Further

The PDE provides a number of templates that allow you to explore many features of plug-in extensions. When creating a new plug-in project, at the **Plug-in Content** dialog, click **Next** instead of **Finish** as we did when we created our first plug-in. This brings up the dialog shown in Figure 7.12. Select one and you'll get a description of what the template creates and what extensions it uses. These are complete, working examples, with reasonably well commented code.

Try them out!

7.6   Rich Client Platform (RCP)

Up to this point we've been using PDE to create plug-ins that operate from within the workbench. We can also use PDE to create so-called *rich client applications* that build on the Eclipse plug-in architecture, but run as independent programs. The Rich Client Platform, first introduced in version 3.0, is basically a refactoring of the fundamental parts of the Eclipse UI that allows it to be used for non-IDE applications.

Select **File –> New –> Project**... **–> Plug-in Project**. Call it "HelloRCP." In the Plug-in Content dialog, answer "Yes" to "Would you like to create a rich client application?" Click **Next** to bring up the Templates dialog of Figure 7.13 and select Hello RCP. Click **Next**. This dialog is just some basic information about our trivial Hello World RCP application. Click **Finish**.

The new project opens in the Manifest editor and looks pretty much like the SuperView project we created earlier. Click the Extensions tab and expand both entries. What makes this an RCP application rather than an Eclipse plug-in is that we're creating an extension to `org.eclipse.core.runtime.applications`. This is the class that

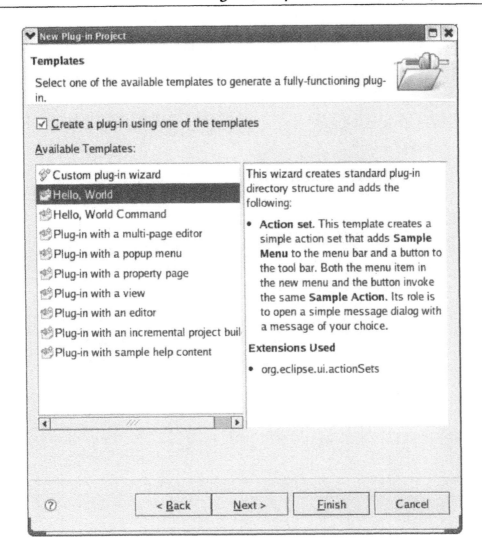

Figure 7.12: PDE plug-in templates.

will be executed when the platform is started. In effect, our plug-in becomes the main application program. We're also creating a new perspective.

Expand the HelloRCP project down to the individual .java files under `hellorcp` and open `Application.java`. This file defines a class that implements `IApplication`, which in turn implements both a start and a stop method.

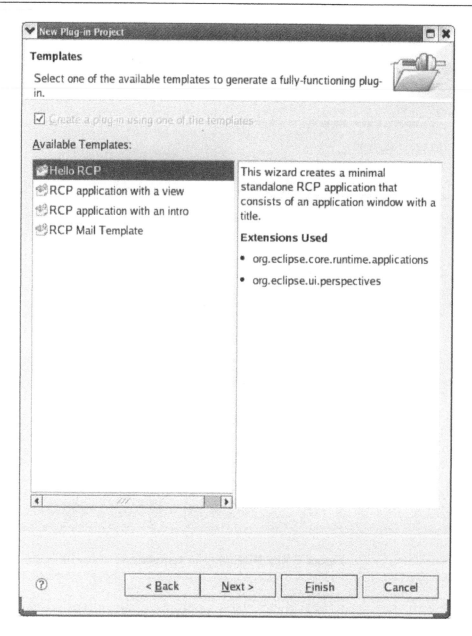

Figure 7.13: RCP Templates dialog.

Run the project by selecting **Launch an Eclipse application** from the Testing section of the Manifest editor Overview tab. The result (Figure 7.14) isn't very exciting, but it does illustrate that we've fired off an independent application window. The code that implements the window is in `ApplicationWorkbenchWindowAdvisor.java`.

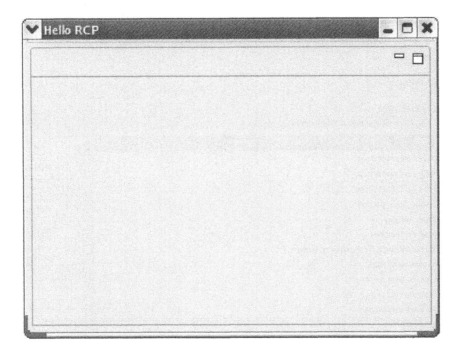

Figure 7.14: HelloRCP application window.

7.6.1   Making It a Product

Ultimately of course, we want to create a stand-alone product that doesn't require Eclipse to run. Right-click the HelloRCP project entry and select **New** –> **Product Configuration**. All you have to do is give it a File name. Call it "Hello."

A new file shows up under the project called `Hello.product` and that file is opened in the Product Configuration editor. Give the product a name, say "Hello RCP." Click the **New**... button to create a new product ID. The default values are fine, so just click **Finish**.

In the Configuration tab click **Add**... and select HelloRCP from the Plug-in Selection list. Then click **Add Required Plug-ins**. The Configuration tab should now look like Figure 7.15.

Go back to the Overview tab, save the file, and click **Synchronize** to get this configuration in sync with the plug-in. At this point it's a good idea to test the application one more time by clicking **Launch an Eclipse application**. If the Hello RCP window came up correctly, then we're ready to export the finished product.

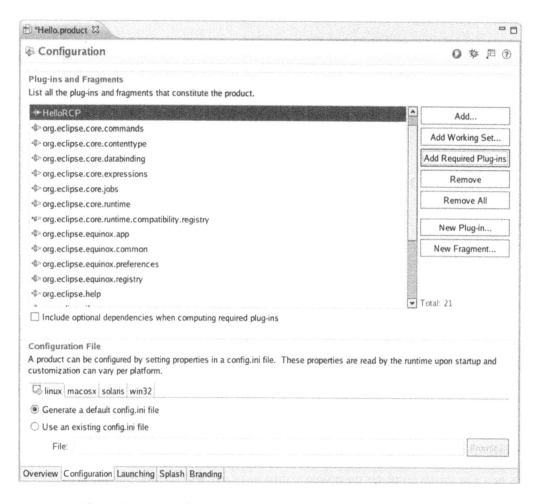

Figure 7.15: Product Configuration editor, Configuration tab.

Click on **Eclipse Product export wizard**. Select a relevant name for Root directory. I called it "HelloApp." We'll store the final application in a Destination Directory. Click **Browse...**, which starts in your workspace. I chose to just put it in `HelloRCP`. Click **Finish** and the final application will be built.

Go to the root directory you just created and you'll see that the executable is called eclipse. You've just created a stand-alone rich client application.

A much more extensive and interesting example of a rich client application is available from the Eclipse CVS repository. Go to the CVS Repository Exploring perspective and add a new repository. The details are:

Host:	dev.eclipse.org
Repository Path:	/cvsroot/eclipse
User:	anonymous
Password:	(leave blank)
Connection Type:	pserver

After connecting, expand the HEAD branch to find a *very large* number of projects. Scroll down to find `org.eclipse.ui.examples.rcp.browser`. Check out that project. Note, incidentally, that there are quite a few `ui.examples` projects. Later on, you might want to investigate some of the others.

Back in the Plug-in Development perspective, right-click the new project entry in Package Explorer and select **PDE Tools –> Open Manifest**. In the Testing section of the manifest Overview page, click **Launch an Eclipse application** to bring up the browser window of Figure 7.16.

This project already has a `Browser.product` file that you can turn into a stand-alone application, as we did above with HelloRCP.

7.6.2   Embedded Rich Client Platform (eRCP)

The RCP is aimed at developing platform-independent desktop applications. So what does that have to do with embedded development, you may ask? The answer is another Eclipse project called the embedded Rich Client Platform (eRCP). This project aims to bring the power of rich client applications to embedded and mobile devices.

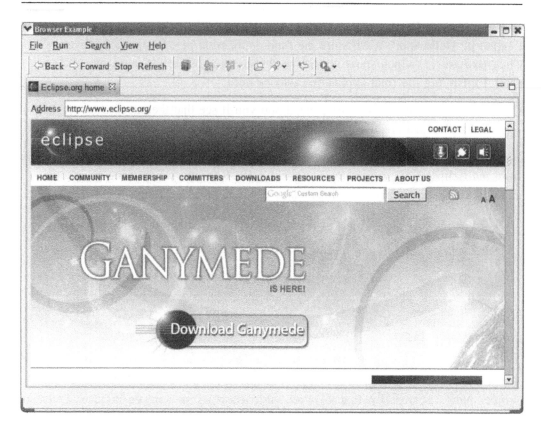

Figure 7.16: Rich Client Platform browser application.

To explain how eRCP differs from RCP, it's necessary to delve a little deeper into the architecture of Eclipse. Referring back to Figure 1.1, the Eclipse workbench is built on top of two hierarchical graphical toolkits. The lower level toolkit is called the Standard Widget Toolkit (SWT), a Java-defined layer sitting on top of platform-dependent GUI components. Every platform on which Eclipse runs has its own native SWT layer, but the Java side of that layer is constant.

The JFace toolkit is built on top of the SWT and is a high-level GUI layer that addresses the needs of Eclipse itself. JFace provides components that create views and support events, control tasks using progress components, and manage UI resources such as fonts. Of course, the GUI requirements of Eclipse are common to many applications, which makes JFace useful outside the Eclipse environment.

eRCP implements a subset of the SWT called, not surprisingly, the embedded Standard Widget Toolkit (eSWT) that provides a set of controls, panels, and other widgets commonly used as building blocks of user interfaces in embedded devices Additionally, eSWT introduces a new component, mobile extensions, primarily targeted at the needs of mobile devices such as PDAs and smart phones.

The design of SWT emphasizes portability among operating systems by keeping the native code layer as small and simple as possible. That's fine in desktop environments where there's ample processor horsepower to compensate for the performance hit of platform-independent code. But in mobile devices, performance is a critical issue. So eSWT sacrifices portability to put more of the functionality in the native layer.

eSWT consists of three elements:

- Core

- Expanded

- Mobile extensions

Figure 7.17 illustrates where eSWT fits in the scheme of things.

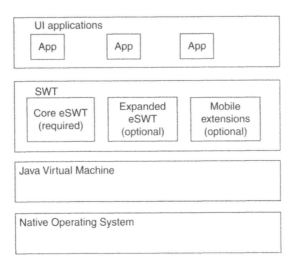

Figure 7.17: eSWT UI toolkit architecture.

eRCP then is a collection of runtime libraries supporting a range of platforms that includes:

- Windows Mobile 2003/5/6

- WindowsCE 5.0 Professional

- Nokia series 80: Includes emulator

- Windows desktop: For testing

Unfortunately, there's no Linux runtime, so in order to play around with eRCP, you'll need a Windows Eclipse installation. Download the latest version of the Windows desktop runtime from the eRCP Download Page at `http://www.eclipse.org/ercp/downloads-page.html`. This is a zip file that creates its own directory. I suggest unzipping it in your Eclipse directory.

The eRCP package includes three fairly simple demos that are started from batch files. Try them out. Unfortunately, the demos don't appear to include source code. There is an example[2] with source code in your `EclipseSamples/` directory in the form of a JAR file that you'll need to import into Eclipse.

Click **File –> Import** ... , expand **Plug-in Development** and select **Plug-ins and Fragments**. This brings up the dialog in Figure 7.18. In the **Import from** section, uncheck **The target platform** and browse to your `EclipseSamples/` directory. In **Import As**, select **Projects with source folders** and click **Next**. Only one plug-in will be found. Select that and **Add** it to the **Plug-ins and Fragments to import** list. Click **Finish**.

Back in the Project Development perspective you'll find a new project in the Package Explorer view. Take a look at some of the Java files and open the manifest to get a feel for what's going on.

However, before you can build and run the example, you must point PDE at the correct target environment. Select **Window –> Preferences –> Plug-in Development –> Target Platform**. Browse to the eRCP/ directory under the eRCP installation directory. Back in the Manifest editor launch the application.

---

[2] This particular example comes from a paper describing the embedded Rich Client Platform posted at IBM's Developer Works, at http://www-128.ibm.com/developerworks/opensource/library/os-ecl-rcp/. Check it out.

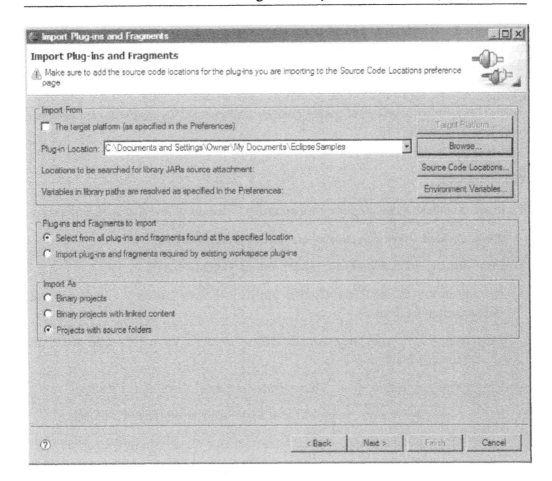

Figure 7.18: Import Plug-ins and Fragments dialog.

Summary

Plug-ins, written in Java, are the mechanism for extending the functionality of Eclipse. The Plug-in Development Environment (PDE), itself a collection of plug-ins, provides intuitive graphical tools to aid the process of plug-in development.

Eclipse defines some 200 extension points where the basic platform can be enhanced by plugging in extensions. The role of a plug-in, then, is to implement extensions.

Normal plug-ins execute as extensions of the Eclipse workbench. The Rich Client Platform (RCP) provides a mechanism that allows you to use the same plug-in development environment to build stand-alone Java applications that utilize Eclipse UI features. Another project, the embedded Rich Client Platform (eRCP), brings the same capability to embedded devices.

The next chapter looks at a couple of advanced features of Eclipse that are also of value to embedded developers, source code control using CVS, and software design modeling using UML.

Resources

There are countless books on Java programming. Here are a few that look interesting.

Block, Joshua. 2008. *Effective Java*. (2nd ed.) Prentice-Hall.

Eckel, Bruce. 2006. *Thinking in Java*. (4th ed.) Prentice Hall.

Horstmann, Cay S., and Gary Cornell. 2007. *Core Java Volume 1—Fundamentals*. (8th ed.) Prentice Hall.

Horstmann, Cay S., and Gary Cornell. 2007. *Core Java Volume 2—Advanced Features*. (8th ed.) Prentice Hall.

Sierra, Kathy, and Bert Bates. 2005. *Head First Java*. (2nd ed.) O'Reilly.

`http://www.java.sun.com/`—This is Sun's website for Java developers.

`http://www.geocities.com/kollurihari/hari/programming.html/`— This is an interesting and very extensive website with tutorials on a wide range of software topics including Java. Nothing on Eclipse yet.

Here is a great starting point for writing plug-ins.

Clayberg, Eric and Dan Rubel, Eclipse: Building Commercial Quality Plug-ins, 2nd edition, Addison-Wesley, 2006.

CHAPTER 8

Eclipse Advanced Features

With a solid background in CDT, it's time to turn our attention to some other tools in the Eclipse workshop that can aid software developers. Specifically, in this chapter we'll look at source code control using CVS and software design modeling using UML.

8.1   UML

UML (Unified Modeling Language) is a mechanism for expressing the constructs and relationships of complex systems, in particular software systems. More specifically, it is a graphical notation that can be used to describe the various models of a software system.

Some of the ways UML is useful include:

- Requirements capture

- Expressing system concepts as classes

- Understanding how objects interact with each other in specific scenarios

- Characterizing the life cycle of objects by identifying the various states to which an object can transition

- Organizing classes into packages and subsystems

- Depicting the deployment of components in a final system

The UML 2.x specifications[1] define 13 types of diagrams that are split into three categories:

- **Structural diagrams**. Identify the objects in the model:
  - Class diagram
  - Component diagram
  - Composite structure diagram
  - Deployment diagram
  - Object diagram
  - Package diagram

- **Behavior diagrams**. Describe what must happen in the system being modeled:
  - Activity diagram
  - State machine diagram
  - Use case diagram

- **Interaction diagrams**. A subset of Behavior diagrams, these emphasize the flow of control and data among the objects in the model:
  - Communication diagram
  - Interaction overview diagram
  - Sequence diagram
  - Timing diagram

Clearly then, the role of an Eclipse UML plug-in is to facilitate creating these diagrams and documenting the properties associated with them. UML is not tied to any specific programming language or software development methodology, although it is very much object-oriented and thus is a natural fit for Java and C++.

---

[1] Version 2.0 was released by the Object Management Group (OMG) in 2003.

This chapter can only skim the surface of UML itself. The objective here is to show how a UML editor works in the context of Eclipse. The Resources section at the end of the chapter lists some resources for UML.

It should be noted that issues of modeling are also being addressed by the Eclipse organization itself through the Eclipse Modeling Framework project (EMF). EMF is described as a modeling framework and code generation facility for building tools and other applications based on a structured data model. For the time being anyway, EMF is primarily oriented toward Java programmers and so may not be particularly useful for embedded developers.

8.1.1 Installing Omondo EclipseUML

Eclipse.org does not itself have a complete UML plug-in. Instead, we'll use a free plug-in offered by Omondo, a UML tool vendor. EclipseUML Free Edition supports all UML 2.1 diagrams and is a good package for exploring what UML is all about.

The EclipseUML download page is:
`http://www.eclipsedownload.com/download_free_eclipse_3.3.html`.
There are three distributions available. For our purposes, the correct one is the EclipseUML Free Installer for Windows and Linux. It's a jar file. Download it to your workstation.

Make sure Eclipse is not running, and in a shell window execute:[2]

`java -jar eclipseUML_E330_2007_freeEdition_3.3.0.v20071210.jar`

This brings up an installation dialog. The first step is to select a language and click **OK**. Click **Next** to exit the welcome screen. The next screen is a little confusing in its description of system requirements. It says that three Eclipse plug-ins are required for EclipseUML and implies that one of them, Graphical Editor Framework, is included. In fact, all three of the required plug-ins are included.

Click **Next** and accept the license terms. The next screen shows your Eclipse installation path. Click **Next** again. This brings up the configuration screen of Figure 8.1. Leave all six items checked. Click **Next** and the installation begins. The installer creates an uninstall script should you wish to remove EclipseUML later.

To confirm that the installation succeeded, select **Window –> Preferences**. UML should show up in the left-hand navigation panel.

---

[2] Be aware, of course, that the file name may change.

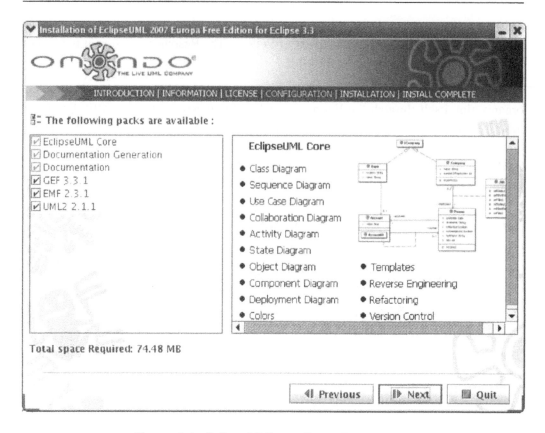

Figure 8.1: EclipseUML configuration screen.

8.1.2   UML Example

Our example for working with UML is a classic embedded control application: an elevator. The problem is to implement the logic required to move the elevator between floors in response to user requests[3]. The elevator operates as follows:

- Each elevator has a set of m buttons, one for each floor. These illuminate when pressed and cause the elevator to move to the corresponding floor. The illumination is canceled when the elevator reaches the corresponding floor.

---

[3] The example was derived from this web page:
http://www.geocities.com/siliconvalley/network/1582/uml-example.htm/.

- Each floor, except the bottom and top floors, has two buttons: one to request an "up" elevator, and one to request a "down" elevator. These buttons illuminate when pressed. The illumination is canceled when the elevator visits the floor and is moving in the requested direction.

- When the elevator has no requests pending, it remains at its current floor with the doors closed.

To begin, we'll need to create a Java project. The reason for starting out in Java is simply that UML tools seem much better integrated with Java. Change to the Java perspective and select **File** –> **New** –> **Java Project**. Name the project "elevator" and accept the defaults. Accept the defaults on the next screen and click **Finish**.

Expand the newly created elevator project and notice that there's an src/ directory under it. Right-click the src/ directory and select **New** –> **Package**. Name it "elevator" as well. The UML diagram files you will be creating can go anywhere. I suggest creating a new directory under the project to store them. With the elevator project selected in the Package Explorer view, select **New** –> **Folder** and call it "models." The project structure should now look like Figure 8.2.

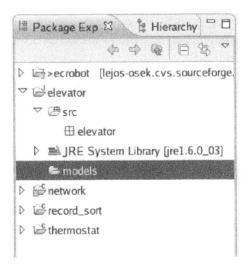

Figure 8.2: Java project structure for UML.

8.1.3   Use Case Diagram

The first UML diagram we'll tackle is the use case. The role of the use case is to describe how an *actor* interacts with the system. It provides a generalized description of how the system will be used. In this case the actor is a passenger desiring to use the elevator.

The elevator system goes through a process something like this:

1. Passenger presses up or down button. Button illuminates.

2. Elevator detects button pressed and moves to passenger's floor.

3. Doors open. Up/down button illumination turns off.

4. Passenger gets in and presses a floor button. Button illuminates.

5. Doors close.

6. Elevator moves to destination floor.

7. Doors open. Floor button illumination turns off.

8. Passenger gets out. Doors close.

Right-click on the elevator package and select **New UML Diagram –> UML Use Case Diagram**. The suggested file name is `elevatorUseCaseDiagram.uud`. I chose to delete the elevator part. Select the `models/` directory and click **Finish** to bring up the Use Case editor. Figure 8.3 shows the editor's toolbar. The icons, starting from the left, are:

- Selection mode

- Zoom mode

- Create an actor

- Create a use case

- Create a system

- Add a generalization

- Add an includes

- Add an extends

- Add an association

- Create a package

- Create an indication

- Create a note

- Create a diagram link

- Create a text label

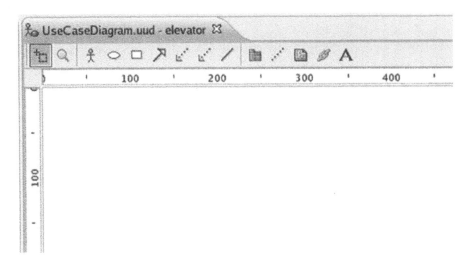

Figure 8.3: Use Case editor tool bar.

Let's start by creating an actor. Click the **Create an actor** icon to bring up the Create an actor dialog. Name it "passenger." The **New Stereotype** button in the dialog is a way of clarifying the role of the actor. For now, leave it blank. Click **OK** and the passenger shows up in the editor window (Figure 8.4).

Next, let's add a use case. Click **Create a use case** and name it "Press up/down button." The use case dialog contains five tabs, the first four of which take free-form text input. In the **Properties** tab, **Abstract** means a use case is not complete, but depends on other use cases. **Pre condition** is the system state before the use case is called, and **Post condition** is its state after this use case is called (Figure 8.5).

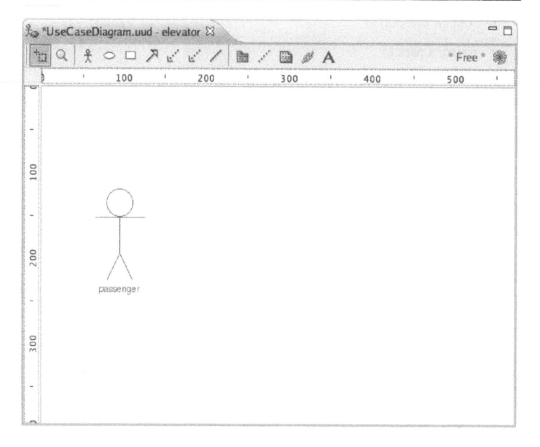

Figure 8.4: The passenger actor.

The **Normal flow** tab is where you describe how the system responds when the use case succeeds. In this case, two activities happen:

- Button illuminates.
- Elevator moves to passenger's floor.

Alternative flow is what happens if something fails. Off hand, I can't think of any actions for that circumstance. **Description** is just additional explanatory text that doesn't fit the other tabs.

The actor and the use case now need to be *associated*. Click the **Add an association** icon, then click both the passenger actor and the Press up/down button use case.

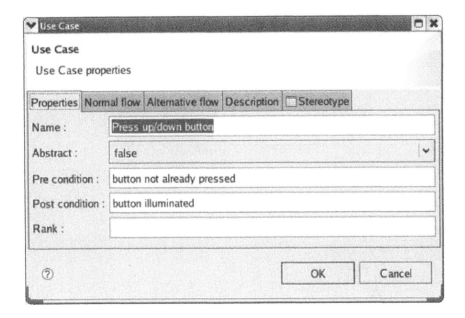

Figure 8.5: New use case dialog.

You can optionally give the association a **Label**. Our use case diagram now looks like Figure 8.6.

We need one more use case for our passenger: "Press floor button." I suggest that the **Pre condition** is that "doors are open" since the passenger can't push the button until he gets in the elevator. The **Post condition** is "floor button illuminated." The normal flow is:

1. Illuminate floor button.

2. Close doors.

3. Move to destination floor.

Create an association between the passenger and the Press floor button case. Finally, we should encapsulate our use cases inside a *system*. Click the **Create a system** button and drag the cursor to create a box around the two use cases. Name it "Elevator." Oddly, the two use case ovals end up outside the system box, so drag them back into it. Figure 8.7 shows the completed diagram. Save it.

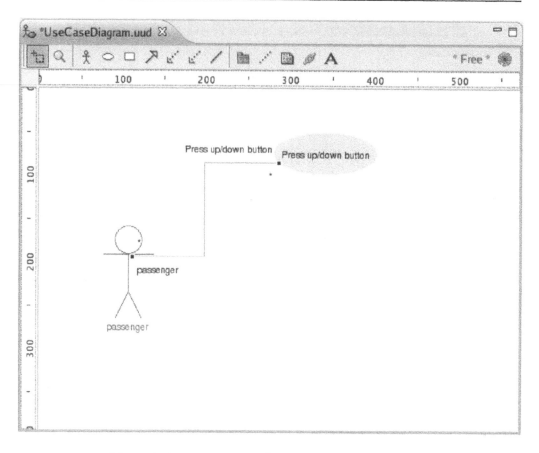

Figure 8.6: passenger and Press up/down use case.

The information you entered into the use case diagram, as well as other diagrams, can be turned into design documentation. Unfortunately, that feature is not supported in the free version of EclipseUML.

8.1.4 Class Diagram

A system, or domain, is composed of classes and the relationships among them. The class diagram depicts these relationships and provides an overview of the domain. EclipseUML lets you define a domain model without getting bogged down in the details of creating classes, interfaces, and relationships. Wizards help you build the diagram

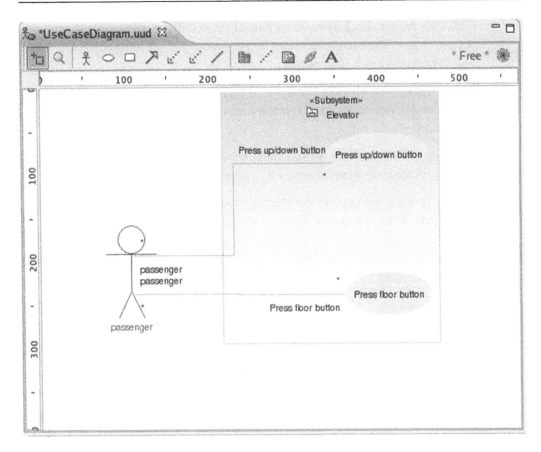

Figure 8.7: Completed use case diagram.

incrementally until you have enough information to adequately describe the problem domain.

For illustration purposes, let's say our elevator has the following classes:

- *Elevator.* The thing that actually moves up and down carrying passengers. The elevator has a current position that might be expressed either as a floor or as an elevation in feet. It can be directed to move up or down from its current position.

- *Door.* The elevator has a door that can be commanded to open or close.

- *Up/Down button*. Each floor has a pair of these buttons, except for the top and bottom floors. The button state can be pressed or not pressed, and it can either be illuminated or not illuminated.

- *Floor button*. The elevator has a set of buttons, one for each floor. Like the Up/Down buttons, these can be pressed or not pressed, illuminated or not illuminated.

- *Controller*. This class ties all the others together by controlling the elevator and door in response to button presses by passengers.

Right-click the elevator package in the Package Explorer view and select **New UML Diagram –> UML Class Diagram**. Change the parent folder to `elevator/models/` and if you choose, delete "elevator" from the file name. The Class Diagram editor toolbar is shown in Figure 8.8. The first two icons are Selection mode and Zoom mode, as in the Use Case editor. The remaining icons are:

- Create a package

- Create a class: drop down menu

- Create an interface

- Create an enumeration

- Association: drop down menu

- Dependency: drop down menu

- Generalization

- Realization: drop down menu

- Create an Element Import: drop down menu

- Create an interface provider

- Create a require interface

- Create an interface connection

- Create a note: drop down menu

- Create an indication

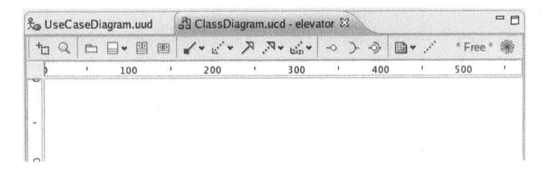

Figure 8.8: Class Diagram editor tool bar.

Click **Create a class** and then click somewhere in the top center of the editor. This brings up the New Java Class dialog. Name the class "Controller" and leave everything else as defaults. Click **Finish**. A class symbol shows up on the diagram. In the Package Explorer view a Java file, `Controller.java`, shows up under the elevator package.

Repeat this process for the following classes:

- Elevator
- Door
- UpDownButton
- FloorButton

The result should look something like Figure 8.9.

Take a quick look at `Controller.java`. Not much there, just a stub for the Controller class. But as we add items to the class diagram, the template code will be expanded to remain in sync with the diagram.

Now that we have the classes, we need to figure out how they interact. Let's start by creating a pair of *directed associations* between the Controller and the two button classes. From the **Association** drop-down menu, select **Directed Association**. Then click on the Controller class. The cursor changes to a plug, implying that we're "plugging" the Controller class into something else. Now click on the UpDownButton class.

Figure 8.9: Class diagram.

A pair of methods show up in the Controller box—getUpDownButton() and setUpDownButton(). This seems reasonable. We want to get the state of the buttons and in turn set the illumination state. Do the same thing with the FloorButton class. Our class diagram now resembles Figure 8.10.

It might be equally useful to create directed associations with the Elevator and Door. Instead, I encourage you to investigate the other associations and connection mechanisms, to see how EclipseUML treats them. And of course, unless you're a Java programmer with some knowledge of UML, they may not make much sense just yet.

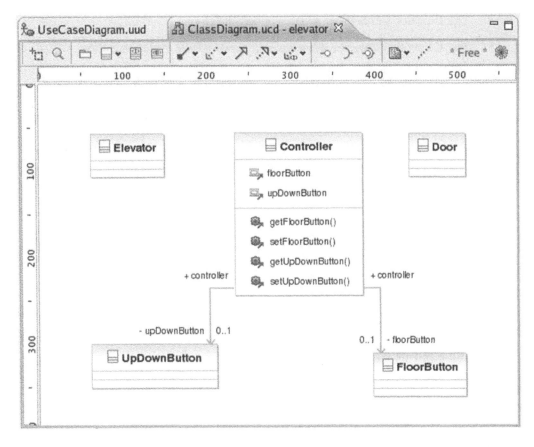

Figure 8.10: Class diagram with associations.

8.1.5 Sequence Diagram

To round out our brief tour of EclipseUML, we'll create a sequence diagram to describe the system's response to the passenger pushing an UpDownButton. The sequence is as follows:

1. The passenger presses an UpDownButton.

2. The UpDownButton sends an update to the Controller.

3. The Controller illuminates the button and moves the Elevator to the specified floor.

4. The Elevator signals that it has reached the floor.

5. The Controller stops the Elevator and turns off the UpDownButton illumination.

6. The Controller opens the Door.

7. After a suitable delay, the Controller closes the Door.

Right-click the elevator package and select **New UML Diagram** –> **UML Sequence Diagram**. Name it "UpDownButtonSequence" and store it in the `models/` directory. The Sequence Diagram editor tool bar (Figure 8.11) has the following buttons:

- Selection mode

- Zoom mode

- Add new property (similar to add an object)

- Create an actor

- Add a message

- Add a self message

- Add a frame

- Add an Interaction Use

- Create a Component

- Create a Class

- Create an Interface

- Create an entity

- Create a boundary

- Create a controller

- Create a note

- Create an indication

- Create a diagram link

- Create a text label

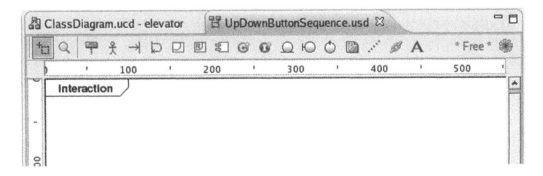

Figure 8.11: Sequence Diagram editor tool bar.

For the purpose of this illustration we'll only be using a couple of these buttons, **Create an actor** and **Add a message**. Our objective here is to express the interactions among the classes in our system for the Press up/down button use case. Begin by creating an actor in the upper left-hand corner of the diagram and name it "passenger."

Now the cool thing is that we can drag the classes from the elevator package in the Package Explorer view directly into the sequence diagram. Drag the classes so they show up in the following order to the right of the passenger:

1. UpDownButton

2. Controller

3. Elevator

4. Door

In order to get the Door class to show up you may have to maximize the editor window and drag the frame out to about 700 pixels. Then restore the editor to its windowed view. Your diagram should now look something like Figure 8.12. Each of the boxes is an instance of its respective class.

Note that the Outline view shows a thumbnail of the entire diagram with the portion visible in the Editor shaded. You can drag the shaded portion to scroll around in the Editor.

In the context of a sequence diagram, classes communicate by sending messages to each other. Click the **Add a message** button, move the cursor to just below the label of the actor (his head turns blue) and click. Now move the cursor to the right until the

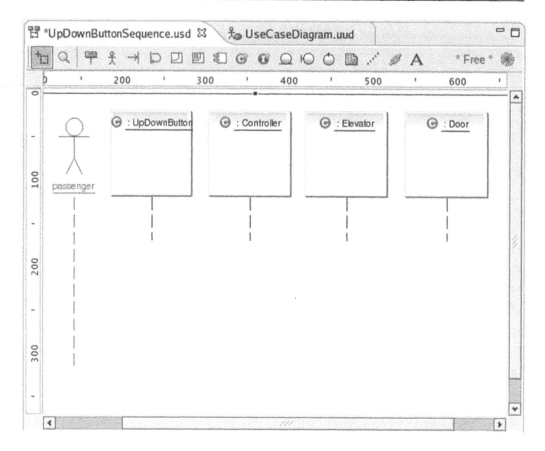

Figure 8.12: Sequence diagram with classes.

UpDownButton box turns blue, and click. The Message dialog of Figure 8.13 comes up. Label it "Press."

Here you can also specify the Operation (method) that carries out the message. In this case, of course, there is none—the passenger is pressing the button. The commercial version of EclipseUML will add method templates to the class code. Leave the rest of the fields at their default values and click **OK**.

In like fashion, add a message labeled "update" from the UpDownButton to the Controller. You might want to grab the vertical box below the controller and drag it down a little to make it clear that the update follows the press. Next add a message from the Controller back to the UpDownButton and label it "illuminate."

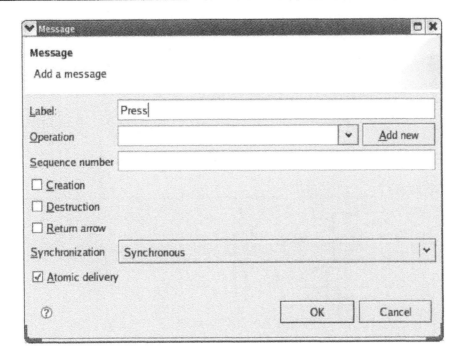

Figure 8.13: Message dialog.

This gets a little tricky. The destination of a message can't be the vertical bar (known as an "activation") below a class instance. Note that when you drag the message wire over the activation bar, the international "no" symbol pops up. Drag the wire until the UpDownButton box is shaded blue, and click. Again, you'll probably want to drag down the activation bar at the end of the illuminate message for readability. Your diagram should now look something like Figure 8.14.

Continue adding messages to carry out the sequence of events listed above. After illuminating the button, the Controller will start the Elevator moving toward the specified floor. The Elevator will notify the Controller when it has reached the floor, whereupon the Controller will stop the Elevator and turn off the button illumination. Finally the Controller opens the Door and, after a suitable delay, closes the Door. The final diagram is shown in Figure 8.15.

Unfortunately, EclipseUML seems to have a mind of its own when it comes to vertical spacing of messages. It seems that some activation bars can be moved and resized, while other can't. The resulting diagram is so big it requires a full screen to display.

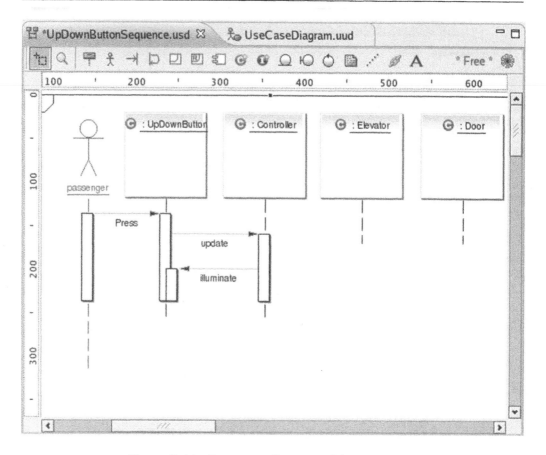

Figure 8.14: Sequence diagram with messages.

We could continue looking at the other UML diagrams, but I think you get the idea. Feel free to play around with them and see what they do.

8.1.6   Configuring EclipseUML

EclipseUML has a number of configuration options. Click **Window –> Preferences** and select the UML entry. The top-level preferences dialog has four tabs. The Appearance tab is show in Figure 8.16. You may very well want to turn off the Show splash screen option. The tool bar can be displayed as text and/or images. The icon images can be either flat or embossed, although I can't really see any difference.

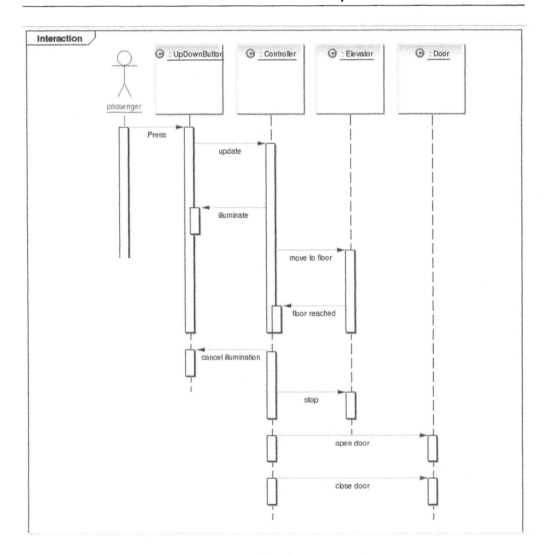

Figure 8.15: Completed sequence diagram.

Diagram presentation style offers some subtle variations in display, mainly involving shading. The differences are more apparent in the use case diagram than in the sequence diagram.

The Options and Print tabs are fairly self-explanatory. The Diagram Board tab offers options related to the grid and ruler, although it's not obvious that they do anything.

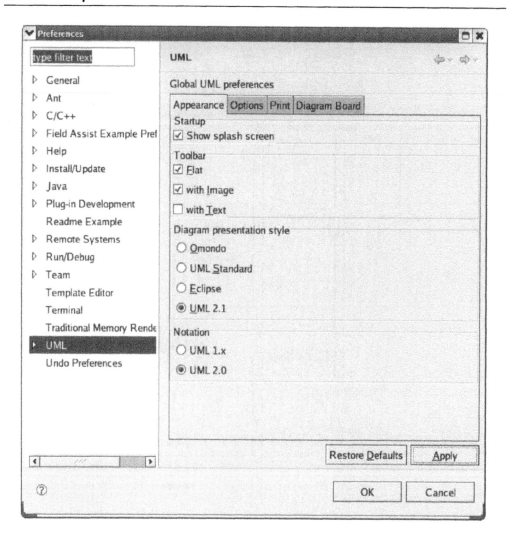

Figure 8.16: EclipseUML preferences.

Expand the UML preferences entry then expand Class Diagram. In the Association dialog is a Router option that specifies how lines, or "wires," are drawn. The default is "Manhattan," which routes the lines with right angles. The alternative is "Manual," which runs the lines directly. Most of the other options are concerned with features of Java.

8.2 CVS

Back in the early days of embedded computing, a lone engineer would design the hardware and write the software, sometimes in assembly language. I did my share of that back in the day. In that kind of environment, keeping track of changes was no big deal. The entire program may have consisted of perhaps a dozen or so files, and it's easy enough to wrap your mind around a project of that magnitude.

Needless to say, things have changed. Projects are not uncommon that contain hundreds, if not thousands of files being worked on by teams of developers who may be distributed all over the world. Now, keeping track of changes is a big deal. In fact, disciplined management of revisions is absolutely critical in maintaining control of the software development process. I even regret not having used version control on projects where I was the only developer.

CVS (Concurrent Versioning System) is an Open Source software package that supports simultaneous development of files by multiple developers. It is commonly used in large programming projects, but its use is not limited to software development. It can be useful in any task that involves managing files of data on a computer system.

CVS uses a client/server paradigm to store a set of files on a server and then make those files accessible to all users who need them. The system provides commands to "check out" a copy of a file for modification and subsequently "commit" changes back to the repository. It also scans files as they are moved to and from the repository, to prevent one person's changes from overwriting another's.

The system also maintains a history of each file, which allows you to go back and recreate any previous version.

8.2.1 Branches

CVS is based on the notion of *branches*, where programming teams can share and integrate ongoing work. A branch is a shared work area that can be updated at any time by any member of the team. This allows individuals to share their own work with other members of the team and to access the work of others during all stages of a project. The branch effectively represents the current shared state of the project.

The process is illustrated graphically in Figure 8.17. Two programmers each check files out of the branch, update them, and commit them back to the branch. It is entirely possible that both programmers are working on the same file. Both programmers need to *synchronize* the file to check for conflicting changes before committing.

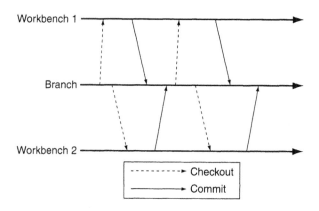

Figure 8.17: CVS workflow.

Every CVS repository has a special branch called HEAD that is the main branch. HEAD, also referred to as the trunk, is considered sacrosanct. You don't commit something to HEAD until you are absolutely certain that it's correct. Other branches are created to provide a safe place to make changes before committing file to the HEAD branch.

8.2.2   CVS in Eclipse

Like most Unix/Linux software packages, CVS is strictly command line-driven. But of course, Eclipse wraps a graphical user interface around that, just like it does for gdb. The CVS GUI is embodied in the CVS Repository Exploring perspective. The blank CVS perspective shown in Figure 8.18 isn't very exciting.

To see how CVS works, we'll have to connect to a *repository* on a CVS server. There's a large number of CVS repositories at `sourceforge.net`. I happened to choose nxtOSEK, an RTOS for the Lego Mindstorm NXT, as my sample project, but feel free to select whatever strikes your fancy at SourceForge, or anywhere else for that matter. nxtOSEK is found at `http://sourceforge.net/projects/lejos-osek`.

In a web browser, go to that page, or the page for your selected project, and scroll down until you find a link to the CVS Repository. Upon clicking the link you'll see

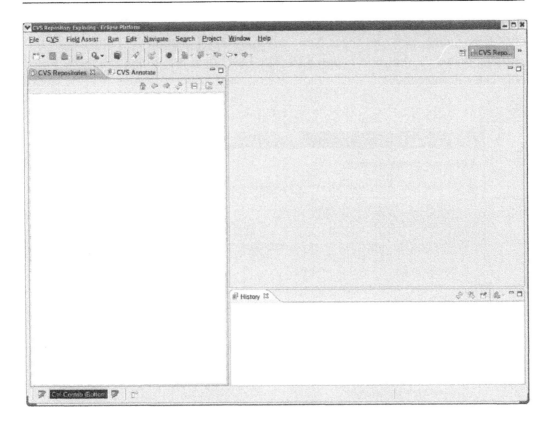

Figure 8.18: Blank CVS Repository Exploring perspective.

some basic information about CVS and about anonymous access. There you'll see a command line, something like:

```
cvs -d:pserver:anonymous@lejos-osek.cvs.sourceforge.net:/cvsroot/
lejos-osek login
```

Note that this is in fact one line.

This tells us the following:

- The repository uses the CVS protocol pserver.
- The user login name is anonymous with no password.
- The server name is `lejos-osek.cvs.sourceforge.net`.
- The root path to the project is `cvsroot/lejos-osek`.

This is enough information to get Eclipse to connect to the project. Right-click in the CVS Repositories view and select **New –> Repository Location** ... Fill out the resulting dialog, as shown in Figure 8.19. The **User:** field is a drop-down menu whose only entry is anonymous.

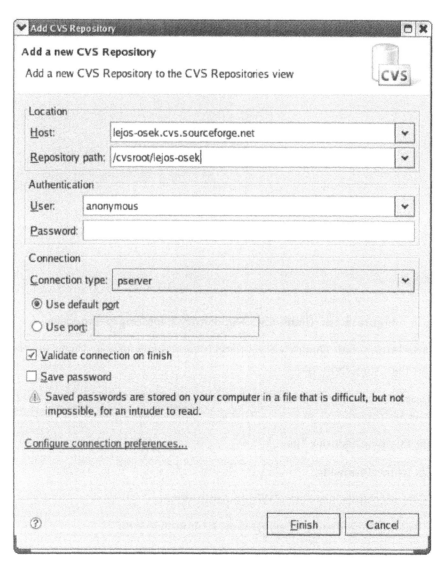

Figure 8.19: Add Repository dialog.

Click **Finish**. You'll find a new entry in the CVS Repositories view. Expand that to look like Figure 8.20. Each time you expand an entry by clicking on the right arrow, Eclipse goes out to the server to retrieve the information for that directory.

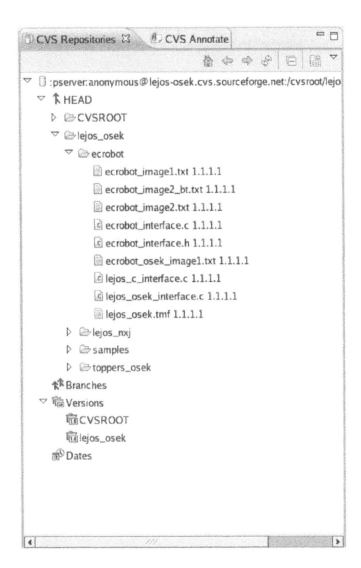

Figure 8.20: CVS Repositories view.

Right-click one of the file entries under the `ecrobot` directory and select **Show History**. The History view (Figure 8.21) now shows the full history of this file including author, timestamp, comments, and so on. You may have to expand the Older than This Month entry to see the history. There's not much here, but it gives you the idea.

Other items in the CVS Repositories view context menu include **Show Annotation** and **Open**, both of which retrieve the file from the repository and open it in a read-only editor window.

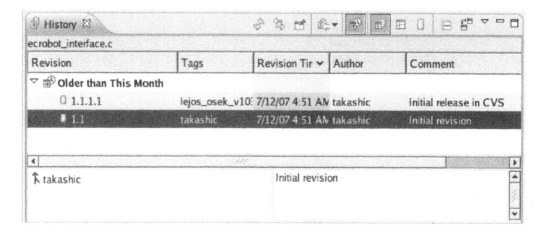

Figure 8.21: CVS History view.

Having connected to a repository, we might want to check out some or all of the contents to work on locally. Right-click the `ecrobot` entry again and select **Check Out**. Eclipse retrieves all the files in `ecrobot` and creates a new project of the same name in the default workspace. The Console view shows the CVS commands and responses.

Go back to the C/C++ perspective and you'll see that the ecrobot project is identified as having come from a CVS repository. All the files have their version number listed, and are identified as ASCII files.

Select **Properties** from the context menu for ecrobot. There's a section called CVS shown in Figure 8.22. This lists all the properties of the repository from which the project came. You can now edit the files in the local project.

The next thing we might want to do is create a branch where we can safely share the project files with other team members, and make changes until we're ready to check them back into HEAD. The branch is created on the repository server and requires write access, which anonymous users normally don't have.

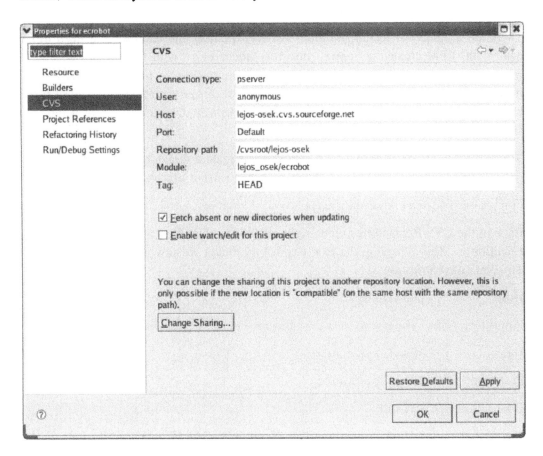

Figure 8.22: Project CVS properties.

8.2.3   Setting Up a CVS server

To continue our exploration of CVS, you'll need write access to a CVS server. If you don't have one, there's a nice server called CVSNT that runs both on Windows and Linux, that's available from `http://www.cvsnt.org/`. CVSNT was originally developed to provide CVS server functionality under Windows and was later ported to Linux. The download for Linux is in the form of a gzipped RPM.

There's a fairly complete installation guide for Linux at the cvsnt.org wiki at `http://www.cvsnt.org/wiki/InstallationLinux`, so I won't bother going into a lot of detail here. You only need to install the required package, `cvsnt-2.5.03.2382-1.i386.rpm`. None of the optional database or protocol packages are necessary for our purposes.

There are a couple of incompatibilities between Eclipse and CVSNT that need to be dealt with. RPM created a `cvsnt/` directory under `etc/` with files `Pserver.example` and `Plugins.example`. Copy `Pserver.example` to `Pserver` and open it in an editor. Find a line that says `#Compat0_OldVersion=0` and uncomment it by deleting the `#`. Then uncomment the line `#Compat0_OldCheckout=0` a little farther down.

You'll need to create an initial CVS repository. Create a directory in a suitable place, `/usr/local/cvsroot` is what I chose, and then as root user execute:

```
cvs -d /usr/local/cvsroot init
```

Back in the CVS Repositories view, right-click and select **New –> Repository Location** … This brings up the Add Repository dialog we saw in Figure 8.19. This time the entries are:

Host: `localhost`

Repository path: `<your_repository_path>` (`/usr/local/cvsroot`)

User: `<your_user_name>`

Password: `<your_password>`

Connection type:`extssh`

The new repository looks something like Figure 8.23. The HEAD and Versions nodes have the requisite CVSROOT entries, but are otherwise empty, as is the Branches node. Our next task then is to add a project to the repository that we can share with other members of our team.

Let's use the thermostat project as an example. Go to the C/C++ perspective and right-click thermostat in the Project Explorer view. Select **Team –> Share Project** … The first dialog box gives you the choice of using an existing repository or creating a new one. Highlight your local repository and click **Next**. Here you have a choice of

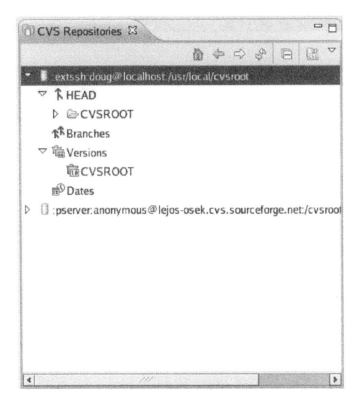

Figure 8.23: New local CVS repository.

module naming options. Keep the default, Use project name as module name. Click **Next**.[4]

The next dialog lets you select which of the project's resources to share (Figure 8.24). Typically you want to share all of them. Click **Finish** to launch the Commit wizard.

Interestingly, the Commit wizard finds a couple of files with "unknown names or extensions" and asks you whether they are binary or ASCII. .cdtproject is in fact ASCII. thermostat_s is binary. Finally, you can enter a comment for the commit operation. Something like "Adding new project" might be appropriate.

---

[4] I encountered an error at this point: "Errors saving CVS synchronization information to disk." CVS created a hidden directory, .settings, in the project workspace that is owned by root and not world-writable. Changing the permissions fixed the problem.

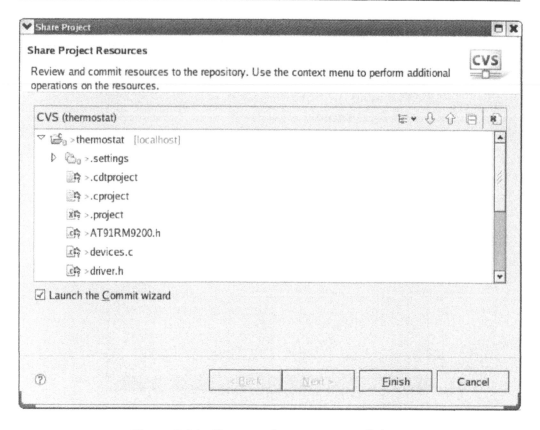

Figure 8.24: Share Project Resources dialog.

The thermostat project is now identified as being under CVS control on the localhost and all the resource files have version numbers. Back in the CVS Repository Exploring perspective, thermostat now shows up as an entry under HEAD in your local repository.

8.2.4   Team Synchronizing

Having committed the project to CVS, we can continue working on our own copy in the workspace and later synchronize any changes with what's in the repository. Make some minor change to thermostat.c, add a comment, perhaps. When you save the changes, greater-than signs (>) appear next to the project name and the file, indicating changes that will need to be checked in to the repository.

Right-click `thermostat.c` in the Project Explorer view and select **Team –> Synchronize with Repository**. This brings up the C Compare Viewer, showing the differences between the Local File on the left and the Remote File in the repository on the right (Figure 8.25). Icons in the tool bar let you step through the differences. You can also copy any non-conflicting changes from the Remote File to the Local File, that is, from right to left.

Where a team of developers is involved, it's quite likely that the file in the repository, the Remote File, contains changes from someone else. The C Compare Viewer highlights those changes as well. In practice, you should perform a **Team –> Update** operation on the file before synchronizing with the repository.

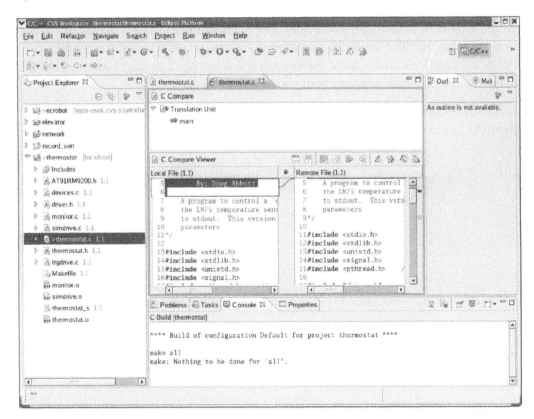

Figure 8.25: C Compare Viewer.

Now you can commit the new version back to the repository. Right-click `thermostat.c` and select **Team –> Commit** ... This operation is also known as "checking in." You're prompted to enter a comment for the commit operation. Then click **Finish**. The new version is written back to the repository.

For completeness, we'll also go through the process of checking a project out of the repository. In the C/C++ perspective, delete the thermostat project. When prompted, check Delete project contents on disk. Now go back to the CVS Repository Exploring perspective and right-click on thermostat. Select **Check Out**. The project is copied to your workspace and built.

8.2.5   Branching

To wrap up our exploration of CVS, we'll look at the branching process. There may be any number of reasons why you would want to create branches for a project in addition to HEAD. Maybe there are multiple versions of a product, in which the software builds differently for each version. Make each one a branch. You might want to establish a developmental branch separate from the production branch.

In the C/C++ perspective, right-click the thermostat project and select **Team –>Branch** ... Give the branch some sensible name (Figure 8.26) and leave the Start

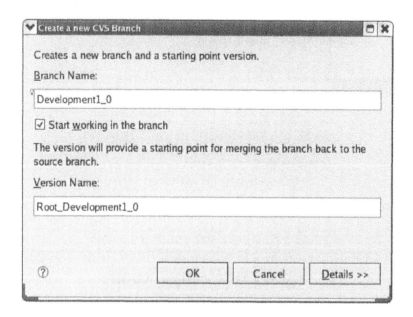

Figure 8.26: Create CVS branch.

working in the branch box checked. CVS proposes a Version Name. The version name identifies the point at which the branch was created, and is necessary later on when you want to merge the branch.

The branch name shows up next to the project name. Back in the CVS Repositories view, expand the Branches node of your local repository. You'll see the branch you just created and under it, the thermostat project. Now, when you do a checkout or a commit, you're working with the newly created branch and not HEAD.

Summary

In this chapter we've looked at two advanced features of Eclipse, one built-in, and the other an add-on, that can greatly improve software productivity and reliability. UML is a powerful tool for visualizing complex software and deriving code templates from the graphical models. Although most open source UML tools tend to be oriented to Java, there's nothing to prevent tools that have a C++ flavor. Eclipse plug-in Central lists 35 plug-ins in its UML category. It might be worth checking some of them out.

CVS is an integral part of Eclipse that brings some order to the potential chaos of team development. Project resources are checked in and out of a central repository in an orderly fashion, and conflicting changes can be easily identified for resolution. CVS tracks every change so that the code can be rolled back to any previous version if necessary.

In the next and final chapter we'll look at how three embedded Linux vendors have adapted the open source Eclipse platform to build commercial products that address various aspects of the embedded development process.

Resources

UML

Fowler, Martin. 2003. *UML Distilled: A brief guide to the Standard Object Modeling Language.* (3rd ed.) Addison-Wesley.

Miles, Russ, and Kim Hamilton. 2006. *Learning UML 2.0.* O'Reilly.

Bruce Powell Douglass has been particularly prolific in writing about UML for embedded and real-time applications. His titles, all published by Addison-Wesley, include:

Douglass, Bruce P. 1999. *Doing hard time: Developing real-time systems with UML, objects, frameworks, and patterns*. Addison Wesley.

Douglass, Bruce P. 2004. *Real-time UML: Advances in the UML for real-time systems*. (3rd ed.) Addison Wesley.

Douglass, Bruce P. 2006. *Real-time UML workshop for embedded systems*. Addison Wesley.

CVS

Bar, Moshe, and Karl Fogel. 2003. *Open source development with CVS*. (3rd ed.) Paraglyph.

Cederqvist, Per. 2002. *Version management with CVS*. Network Theory Ltd.

Vesperman, Jennifer. 2006. *Essential CVS*. O'Reilly.

CHAPTER 9

Eclipse-Based Development Products

An interesting feature of the Eclipse ecosystem is that it explicitly encourages the development of proprietary commercial products on top of the base platform. In the embedded Linux world, several vendors have migrated from proprietary development environments to tools based on Eclipse.

This chapter takes a look at how three of the leading embedded Linux vendors have extended Eclipse with value-added software to create their own unique products. The information is based on evaluation versions of the tools, and for the most part are my personal impressions of what seemed most interesting about each one. This is in no way intended to be a "competitive analysis," or to say that one product is "better" than another. The intention is simply to show how these three vendors have approached the issue.

9.1   Why Buy It?

You may be wondering, why should I pay money for something I can get for free? The most common reason is because open source software vendors have gone through an extensive testing and integration process to make sure that the various software packages from different sources actually do work together as advertised.

The goal of most open source projects is not to create production-ready code, but to push the technology envelope. That's less true of Eclipse with its annual major release of the platform and a large number of related projects. This is, in turn, accompanied by quarterly maintenance releases. The result is code that is quite stable, and for the most part, bug free.

But of course, embedded software development is more than just an IDE. You also need an operating system, tool chain, boot loader, file system, and so on. The quality and stability of open source implementations of these elements varies widely. The Linux kernel, for example, changes daily. Unless you want to be a kernel developer yourself, trying to keep up with that is futile and counter-productive.

The vendors discussed in this chapter offer complete embedded development tool suites that include, at a minimum, the Eclipse platform with proprietary plug-ins, one or more GNU tool chains for cross development, a Linux kernel, and a boot loader of some form. This doesn't come cheap, but if your objective is to get a stable, reliable product out the door on time, it's probably worth it.

9.2 LynuxWorks—Luminosity

Having been in business for some 20-odd years, LynuxWorks predates the Linux craze, selling its own Unix-like real-time operating system called LynxOS. Variants of LynxOS include:

- *LynxOS-SE.* Based on a virtual machine architecture that supports "medium robustness" security as defined by the US Government.

- *LynxOS-Secure.* A separation kernel and embedded system hypervisor for very high security, mission-critical applications

- *LynxOS-178.* Meets the requirements of DO-178B Level A for safety-critical systems

LynuxWorks also offers its own version of Linux called BlueCat Linux.

Luminosity, the company's Eclipse-based IDE, supports the full range of operating system products. The current version of Luminosity, 3.0, is based on Eclipse version 3.2.

9.2.1 Getting the Evaluation

The product page for Luminosity on the LynuxWorks website doesn't offer a link for an evaluation, but the RTOS product pages do offer such a link. This takes you to a request form. The RTOS evaluation will include Luminosity.

My evaluation version of Luminosity was delivered as four ISO image files:

- FlexLM license manager

- Luminosity

- LynxOS cdk

- LynxOS ode

9.2.2   Getting Started

The first step is to install the FlexLM license manager and have it retrieve a host ID number. This turns out to be the MAC address of your eth0 port. Send that to LynuxWorks and they will send back a license file with limited-time licenses for both Luminosity and LynxOS. Before starting Luminosity, you must start the FlexLM daemon.

Actually, you can run Luminosity without a license, in which case it simply reverts to standard Eclipse 3.2. Figure 9.1 is the initial perspective for Luminosity. There are several additional icons in the tool bar, most of which represent items in the LynuxWorks menu. Note also a set of shortcuts for selecting perspectives.

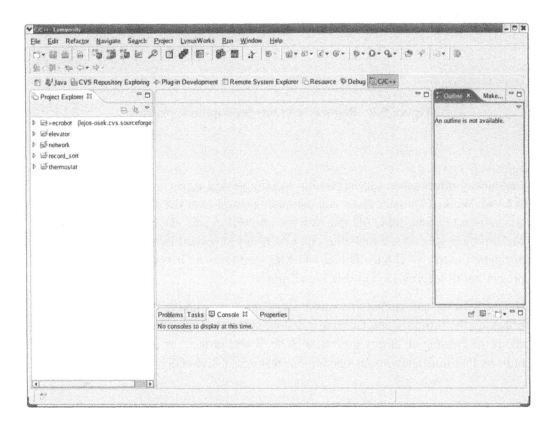

Figure 9.1: Luminosity initial perspective.

In order to create and run embedded projects with Luminosity, you must register a cross-development platform (Figure 9.2). In this case it's LynxOS for the x86.

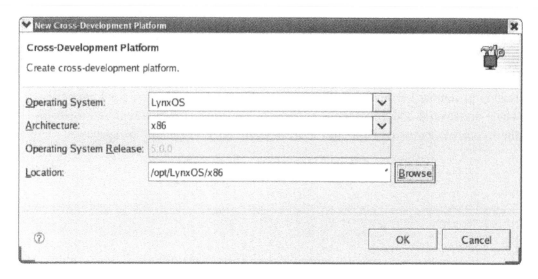

Figure 9.2: Register a cross-development platform.

Luminosity offers several LynuxWorks-specific project types, as shown in Figure 9.3. A LynuxWorks C project gives you complete control over the Makefile. Luminosity generates a template Makefile that you are allowed to edit. By contrast, the Managed Make project generates a makefile that you're not supposed to edit. The same distinction is true of Device Driver and Managed Device Driver projects. The Kernel project builds a LynxOS bootable kernel image.

Having selected and named a LynuxWorks C project, clicking **Next** a couple of times brings up the Project Code Generator dialog, shown in Figure 9.4. Here you have the choice of creating an empty project, a Hello World project, or one of several sample projects that illustrate various operational features of LynxOS.

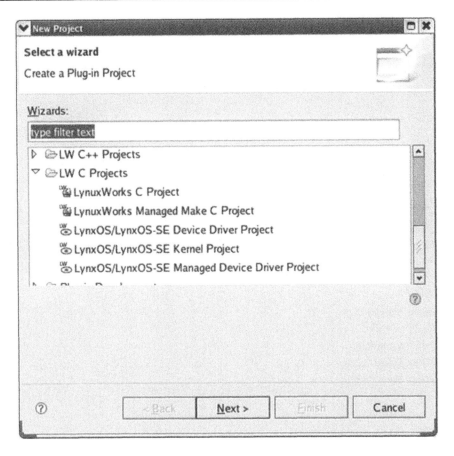

Figure 9.3: New Project dialog and wizard selection.

Clicking Finish creates the project and brings up the LW C/C++ perspective.
This is very much like the standard C/C++ perspective with the addition of a LW
C/C++ Projects view and a LW Make Targets view. The Projects view is similar
to the standard Navigator view with a slightly different look. LW Make Targets
seems to take the place of the standard Make Targets view.

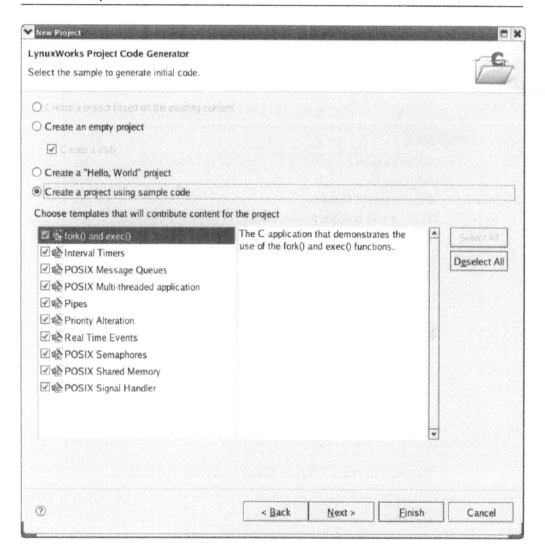

Figure 9.4: Code Generator dialog.

9.2.3 Debugging With Luminosity

LynuxWorks doesn't support any kind of simulator for LynxOS, so you need a real target running LynxOS in order to test code. That's the bad news. The good news is that just about any old 486 box you have lying around as a doorstop should work as a target. There is a tool to build a bootable CD with a LynxOS kernel image.

Luminosity uses a proprietary method for managing remote targets. Select **LynuxWorks –> Set Remote Target** to bring up the Targets view, which is initially empty. Right-click in the Targets view and select **New Target**. This brings up the dialog of Figure 9.5. The Connection tab is fairly self-explanatory. Once you enter the parameters, the Validate button will attempt to resolve the target IP address and then check all connection types that Luminosity uses.

Figure 9.5: Remote Target Configuration.

The Authentication tab lets you enter a user name and password, and a directory on the target where application binaries will be downloaded. The Utilities tab sets network protocol parameters and can usually be left at default values. If you only define one target, it becomes the default.

Luminosity defines its own launch configuration for LynxWorks C/C++ projects. The only tab that differs from the standard C/C++ Local Application launch configuration is the Debugger tab (Figure 9.6). This sets parameters for remote debugging on the default target.

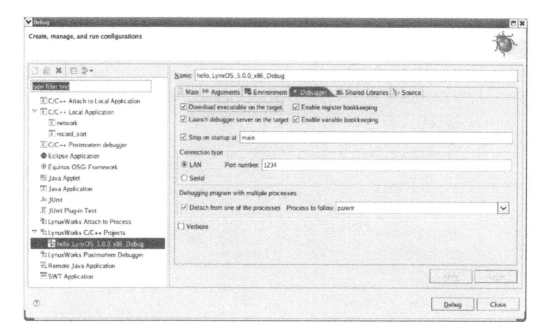

Figure 9.6: Launch configuration debugger tab.

Basic debugging is essentially the same as in standard Eclipse. LynuxWorks does add a couple of its own interesting debug features.

Tracepoints

There are times when you don't really want to, or can't, stop the program at the breakpoint, but you would like to monitor its behavior by watching the value of selected variables. Tracepoints are set in the program in much the same way as are breakpoints. When program execution encounters a tracepoint, any variables or expressions attached to it are evaluated and saved. Later, when the program is stopped, you can review the saved data.

Two new views are associated with tracepoints: Tracepoints and Trace Data. The Tracepoints view is where you define and configure tracepoints. Figure 9.7 shows the Tracepoint properties dialog where you can add expressions to be monitored.

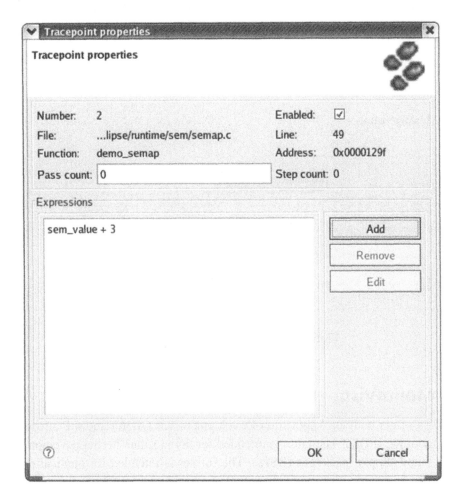

Figure 9.7: Tracepoint properties.

The Trace Data view displays collected trace data when the program is stopped. It provides menu items to start and stop tracing, scroll through the collected data, and save the data to a file.

POSIX IPC Awareness

Luminosity can display information about POSIX inter-process communication mechanisms being used by the application being debugged. This information is contained in a set of views representing each of the IPC mechanisms:

- Semaphores

- Mutexes

- Conditional variables

- Message queues

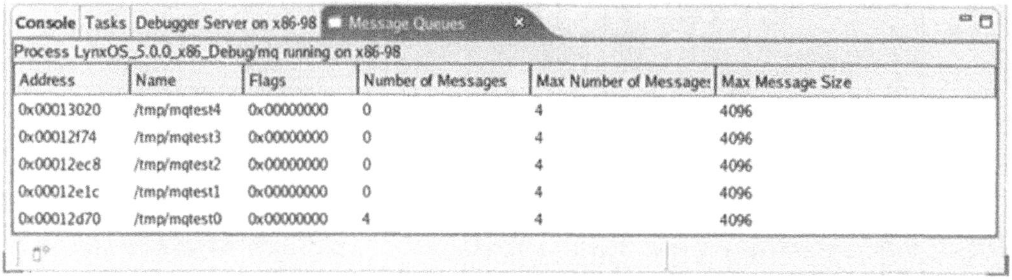

Address	Name	Flags	Number of Messages	Max Number of Messages	Max Message Size
0x00013020	/tmp/mqtest4	0x00000000	0	4	4096
0x00012f74	/tmp/mqtest3	0x00000000	0	4	4096
0x00012ec8	/tmp/mqtest2	0x00000000	0	4	4096
0x00012e1c	/tmp/mqtest1	0x00000000	0	4	4096
0x00012d70	/tmp/mqtest0	0x00000000	4	4	4096

Figure 9.8: Message Queues view.

9.3   MontaVista—DevRocket

MontaVista offers both an Application Development Kit (ADK) and a Platform Development Kit (PDK). The latter is intended for doing Linux kernel development and building board support packages (BSPs). The former is intended for application development and leaves out the features that support kernel and BSP development. Both packages are based on the company's DevRocket IDE, which in turn is built on Eclipse.

9.3.1   Getting the Evaluation

For purposes of this book, I tested an on-line demo of DevRocket that includes a simulation of a Power PC target board. MontaVista's website includes an "Evaluation

Center," accessed from the Products and Services tab of the main page. From there you can fill out a form to request access to the online demo. You should then receive an email with a link to the demo, a user ID, and a password. Access is time-limited. In my case it was about two weeks.

What's Included

ADK version 5.0 includes:

- Eclipse version 3.2

- CDT version 3.1

- Remote System Explorer 1.0

- GNU cross tool chain for a specific architecture, based on GCC 4.2

- Analysis and optimization tools:

  - Application pre-linking

  - Library optimization

  - Memory leak detection

  - Memory usage analysis

  - Application profiling

  - Linux Trace Toolkit: in PDK only

- Linux kernel version 2.6.18

- "Virtual target" based on VMware for testing without actual target hardware

9.3.2 Getting Started

After obtaining a user ID and password from MontaVista, you are directed to a specific web page from where you can start the simulation. This brings up a Gnome graphical desktop, as shown in Figure 9.9.

The desktop initially has three windows open: DevRocket itself, an Xterm serial console connected to the simulated target board, and the target simulation. The first

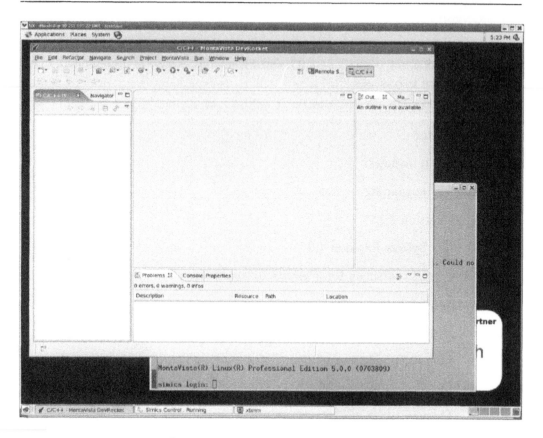

Figure 9.9: DevRocket simulation.

thing to notice is a new top-level menu item, **MontaVista**, the contents of which are shown in Figure 9.10.

This menu provides access to most of MontaVista's extensions. Note the range of new objects that can be created. The memory analysis and profiling tools can be started from here. There are facilities for license and edition management that bring up corresponding preferences pages. The **RSS Feeds** item brings up a Feeds view where you can see RSS feeds from your favorite news sites.

This version includes the Platform Development features so the menu includes a number of kernel development items. The **Kernel Project** lets you configure and build a Linux kernel and subsequently debug it using KGDB.

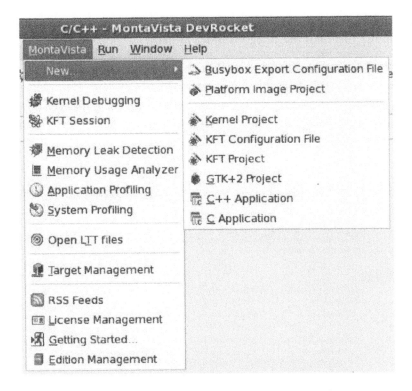

Figure 9.10: MontaVista menu.

KFT stands for Kernel Function Trace, which is effectively a profiling tool for the kernel. It adds instrumentation callouts to every kernel function entry and exit to generate a trace log of function execution with timing details. This adds considerable overhead to the kernel, so it's not particularly good for revealing precise timing problems such as race conditions, but it is useful for identifying bottlenecks such as functions with long execution times and those that are called frequently.

9.3.3 Platform Image Builder (PIB)

One of the tasks required for an embedded Linux device is to create a file system. What goes in it, what can you leave out? Creating a file system by hand can be a tedious, iterative process as you work through the various feature dependencies.

Platform Image Builder is a combination of a project creation wizard and a perspective to help guide you through the process of building a file system from RPM packages.

You can also add your own applications and libraries from projects in your workspace by importing the files into the Image Builder project.

The Platform Image perspective, Figure 9.11, includes a *Platform Image Builder editor* that lists the packages available for inclusion in the file system. Here you select the pre-defined packages you need and the wizard works out the dependencies from the RPM database as you select them. The editor includes filtering and grouping options to make it easier to navigate through the package list.

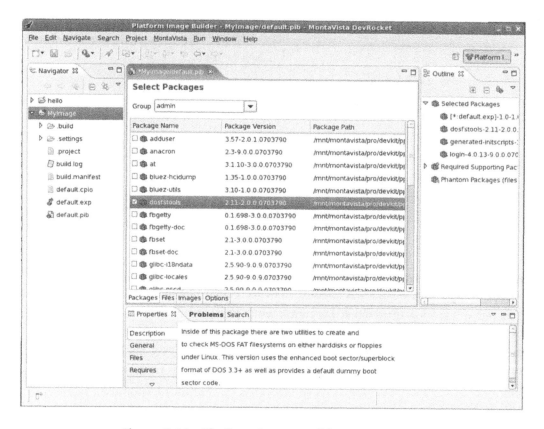

Figure 9.11: Platform Image Builder perspective.

When you highlight a package in the **Packages** tab, details about the package such as version, release information, and dependencies show up in the Properties view. This information comes directly from the RPM database. Other tabs in the editor include

Files, which details the structure of the file system, and **Images,** which lets you specify one or more mount points on the target. Each mount point gets its own image. The **Options** tab lists optional features of the build. The output of the Platform Image Builder editor is a file called `default.pib`.

There's another file in the project called `default.exp`, which stands for "exports." The name strikes me as a little odd because what it really does is identify directories, files, and symbolic links to be added to the file system. This file is managed by an *Exports editor.*

The Outline view shows the packages that have been selected as well as the required supporting packages.

The final output of an Image Builder project is a binary image of the file system called `default.<file_system_type>` where `<file_system_type>` represents the type of file system you're building, such as `ext2` or `ext3`, `jffs2`, and so on. The file system type is specified in the Images tab of the Platform Image Builder editor.

9.3.4   Memory Analysis Tools

DevRocket includes two memory analysis tools: a memory leak detector and a usage analyzer. The leak detector is based on mpatrol, an open source library that wraps `malloc()` and `free()` functions, and the C++ `new()` operator, with instrumentation that logs each call to these functions. mpatrol is a dynamically-linked shared library, so there's no change to the application.

DevRocket wraps the text-based mpatrol library and utilities with a graphical Eclipse front end to make it easier to work with. To run a program with memory leak detection, you create a Memory Leak Detection run configuration from the **MontaVista** menu. This is much like any other run or debug launch configuration, with the addition of a tab for configuring mpatrol (Figure 9.12). The defaults seem to be just fine.

When you click **Run,** DevRocket invokes mpatrol to run the program and then brings up the Memory Leak Detection perspective, with views that display the contents of the mpatrol logs. Figure 9.13 shows a Call Graph in the Memory Leak Detection view.

The memory analysis tool is intended to help you find memory usage problems quickly and accurately by providing a graphical depiction of memory usage across the entire

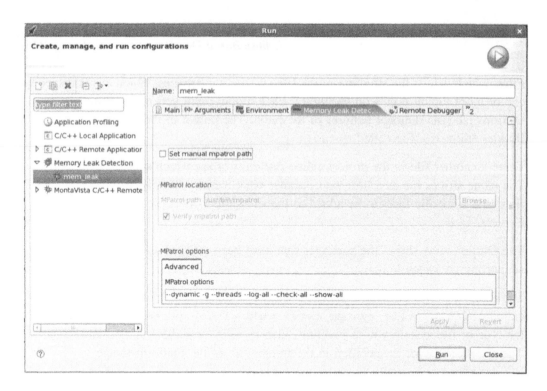

Figure 9.12: Memory leak detection launch configuration.

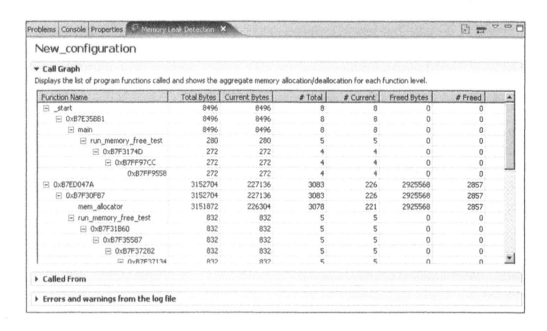

Figure 9.13: Memory Leak Detection view.

system. Memory analysis starts with a high-level view showing relative memory usage for the whole system. From there you can drill down into kernel and application-specific memory usage. Drilling down even further, you can view a memory map for each application.

From the menu bar, select **MontaVista –> Memory Usage Analyzer**. This brings up the Memory Usage Analyzer view, shown in Figure 9.14. From here you can click on Applications or Kernel to get more detail. Figure 9.15 shows memory usage by user space applications. The kernel usage graph shows memory allocated by `vmalloc()`, allocated as slabs, and memory used by page tables.

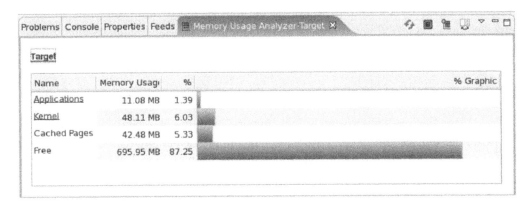

Figure 9.14: Memory Usage Analyzer view, system level.

| Problems | Console | Properties | Feeds | Memory Usage Analyzer-Target ✕ |

Target

Name	Memory Usage	%	% Graphic
Applications	11.08 MB	1.39	
Kernel	48.11 MB	6.03	
Cached Pages	42.48 MB	5.33	
Free	695.95 MB	87.25	

| Problems | Console | Properties | Feeds | Memory Usage Analyzer-Target ✕ |

Target > Applications

Name	Pid	Total Size	Resident Size	Effective Size	% Effective	% Effective Graphic
sshd	660	7.32 MB	5.13 MB	3.67 MB	33.14	
sshd	195	4.57 MB	3.6 MB	1.5 MB	13.57	
sftp-server	664	4.09 MB	3.38 MB	1.27 MB	11.42	
bash	615	2.96 MB	2.67 MB	1.21 MB	10.88	
dhclient	159	2.46 MB	2.2 MB	1.17 MB	10.59	
bash	3333	2.77 MB	2.53 MB	070.05 KB	0.55	

Figure 9.15: Memory Usage Analyzer view, applications.

The information for the Memory Usage Analyzer comes from /proc files. There are a number of virtual files in the /proc directory and in the subdirectories for each process that provide information on memory usage. The Memory Usage Analyzer simply collates this information in a convenient, easy-to-understand format.

9.4   Wind River—Workbench

Wind River's entry in the Eclipse IDE sweepstakes is called, simply, Workbench. The current version is 3.0, based on Eclipse version 3.3.1. Workbench supports both Linux and Wind River's proprietary operating system, VxWorks. In addition to the standard GNU compiler, Wind River also supplies its own compiler with support for multiple target architectures. In addition to supporting both VxWorks and Linux code development, the Wind River compiler also supports stand-alone applications. It also has a number of simulators so you don't need a target for debugging.

9.4.1   Getting the Evaluation

I downloaded an evaluation version that supports on-chip debugging. Workbench for On-Chip Debugging (OCD) uses in-circuit emulation (ICE) to target tasks such as board bring-up and flash programming, where a software debugger such as GDB may not yet be operational. While the full Workbench OCD product supports debugging Linux and VxWorks projects, the evaluation version is limited to stand-alone projects.

From the **Products** tab on Wind River's web site, select **Download Center**. Then select **Software**. Listed among the top three evaluations is **Workbench 3.0 for On-Chip Debugging**. Click that link, then click **Download**, and you're given the choice of downloading or getting a CD. Clicking **Download** here takes you to an information form to fill out. After completing the form you'll be able to download the 600 MB file.

Wind River uses the FlexLM license manager. The evaluation comes with a 30-day license.

9.4.2   Getting Started

When you start this version of Workbench, after getting past the workspace selection and Welcome screen, you're presented with a start dialog that offers choices such as creating or editing a launch configuration, connecting, or syncing with a target. The main editor window shows a list of Getting Started Resources (Figure 9.16) with links to documentation and other support resources. Workbench supports an RSS Feeds view that is already subscribed to several Wind River feeds. Roll the cursor over an item in the Feeds view and a pop-up appears, displaying a summary of that item.

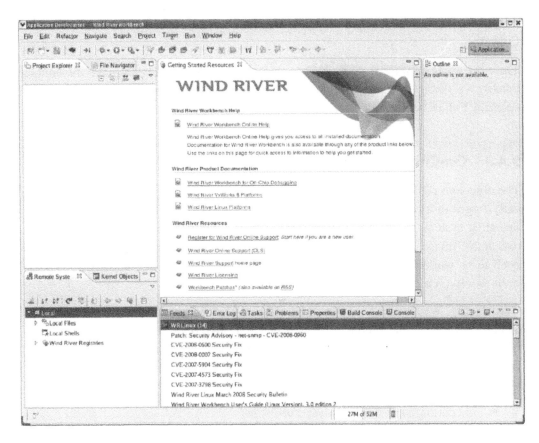

Figure 9.16: Workbench initial screen.

The default perspective is called Application Development. It's similar to the C/C++ perspective with the addition of several views, including:

- Remote Systems
- File Navigator

- Kernel Objects

- Feeds

- Build Console

Workbench makes a distinction between consoles that are attached to targets and the local console that displays the build results. The latter is called the Build Console.

There are some additional items in the tool bar. There's a feature to make the editor emulate vi. Another feature turns on Emacs key bindings, replacing some of the standard Eclipse key bindings.

9.4.3   On-Chip Debugging

The objective of the OCD version of Workbench is to provide tools that help in the early stages of a project, such as bringing up a new board. At this point there's no operating system, so OCD supports a stand-alone application environment where the executable image is expected to be self-contained. We also can't expect the hardware itself to be working, so usually a hardware debugging device such as a JTAG probe or an in-circuit emulator (ICE) is called for.

Accordingly, Workbench OCD provides Remote System connections for two classes of hardware debugging products, Wind River ICE and Wind River Probe. Both products use either JTAG or BDM ports to connect to a target board. Probe connects to the host through USB, while ICE connects via Ethernet and an RS-232 port. Both devices support typical debugging features, such as:

- Execution control: download, start, stop

- Breakpoints

- Examine and modify memory and registers

- Flash programming

- Hardware Diagnostics

Additionally, the ICE class products offer profile analysis and trace capabilities.

Wind River provides a number of sample programs to illustrate various features of Workbench. Under **File –> New** is an **Example** ... menu item that can be used to create various sample projects. One class of examples is that of stand-alone projects that require no OS support (Figure 9.17). The C Demonstration program is a good starting point.

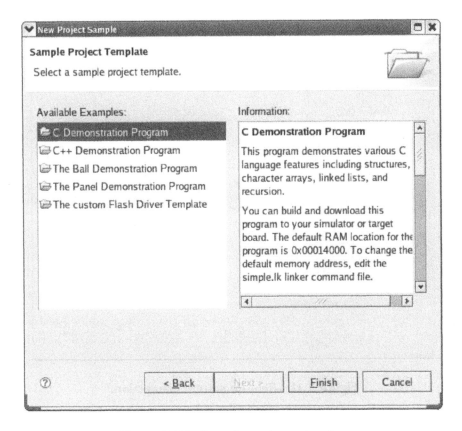

Figure 9.17: Sample project selection.

In place of an actual ICE or Probe device, the evaluation version provides an instruction set simulator for a PowerPC MPC8260 processor. You establish a connection to the instruction set simulator in the same way you would to any other type of remote system (see Figure 9.18).

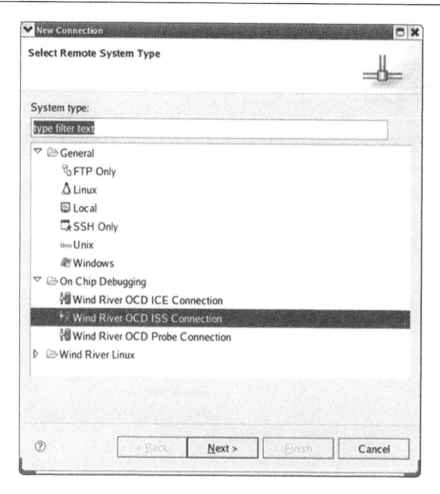

Figure 9.18: Remote system connection.

Next, you create a launch configuration for the project that uses the ISS connection. The dialog for setting up an OCD launch configuration is somewhat different from the launch configurations we've seen before (Figure 9.19). Perhaps not surprisingly, the dialog deals with a lot of hardware issues. The Download tab lets you specify the files that will be downloaded to the target.

Clicking **Debug** brings up the Device Debug perspective, which is similar to the standard Debug perspective, with some additional views (Figure 9.20). A System Context view appears in the editor window. This is a mixed C and assembly language listing of the code being executed. The Debug Symbol Browser shows all of the

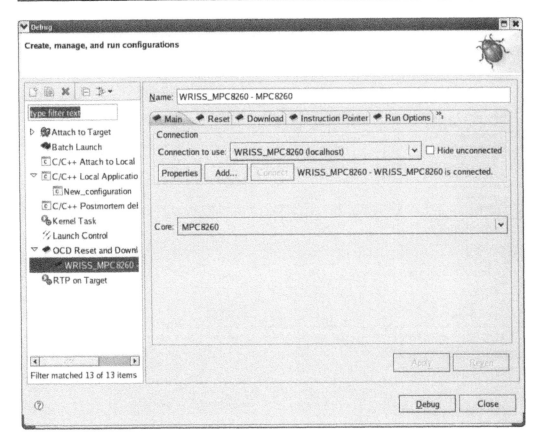

Figure 9.19: OCD launch configuration.

symbols in the project and supports operations such as going directly to the symbol's declaration and setting a breakpoint at the symbol.

From this point on, debugging is essentially the same as we've seen earlier. Workbench OCD supports some additional views that are only meaningful if the debugging tool provides the necessary data. These include:

- Flash Programmer

- Hardware Diagnostics

- OCD Statistical Code Profiling

- Trace

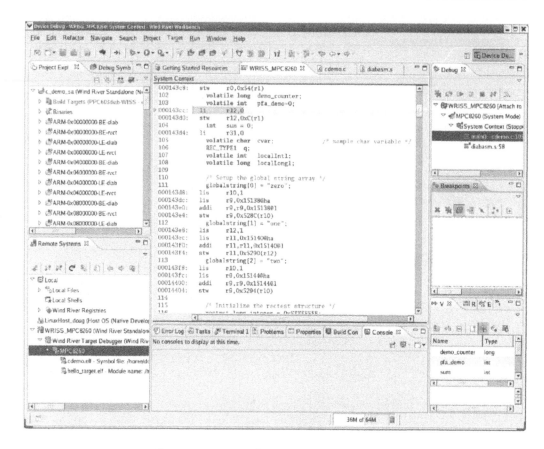

Figure 9.20: Device Debug perspective.

9.4.4   Analysis Tools

The full Workbench product includes several analysis and visualization tools similar to what we've encountered with the other products reviewed in this chapter. They aren't included with the Workbench OCD evaluation.

All but one of these tools (System Viewer) were previously offered as an add-on product called Scope Tools for Test and Validation. These are now included in the base Workbench platform. Each of these tools has its own Eclipse perspective with views appropriate to the task at hand.

System Viewer

This is the Wind River equivalent of the Linux Trace Toolkit. It was previously
called WindView and has been around for quite some time[1]. Like LTT it provides a
graphical visualization of system events to reveal the complex interactions of tasks,
threads, and interrupts (Figure 9.21). I think of this kind of tool as a logic analyzer
for the software.

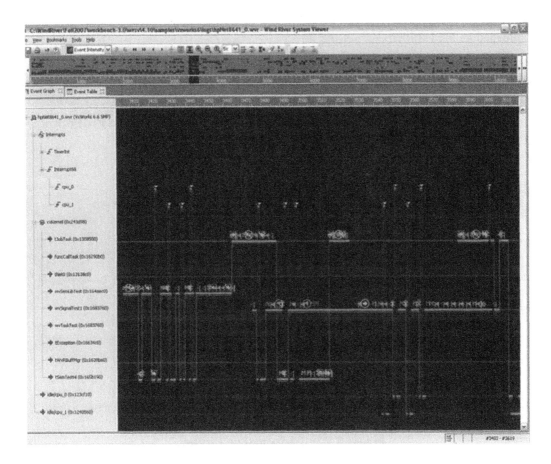

Figure 9.21: System Viewer.

---

[1] I remember playing with an early version with VxWorks back in the mid-1990s.

You can scroll around the trace and zoom in on particular areas to see more detail. The set of processes, threads, and events being traced can be filtered to focus attention on specific areas of concern.

Performance Profiler (Formerly ProfileScope)

This is a dynamic performance profiler that shows where a program is spending its time. An agent on the target periodically takes a "snapshot" of the currently executing process and its call stack. These snapshots are saved in a buffer and periodically uploaded to the Profiler GUI on the host.

The Profiler graphically reports the percentage of CPU time spent in any function (Figure 9.22). This view organizes the information in a call stack format. Current Direct % is the time spent in the function itself. Current Indirect % is the time spent in the function and all functions that it calls. Click on a function name in the Performance Profiler, and an editor opens at that function.

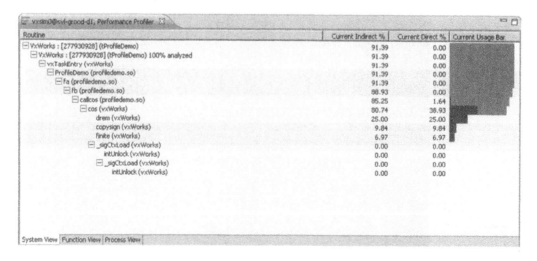

Figure 9.22: Performance Profiler.

Memory Analyzer (Formerly MemScope)

This tool serves as a memory leak detector much like MontaVista's implementation of mpatrol. On the target it dynamically patches the memory allocation functions with instrumentation code. This approach means the application doesn't have to be rebuilt for memory analysis and also means you can analyze any code, not just your own.

Like the Performance Profiler, the data is collected in a local buffer and periodically uploaded to the Memory Analyzer GUI on the host.

Data Monitor (Formerly StethoScope)

The Data Monitor uses an oscilloscope metaphor to monitor program variables in real time and display the results graphically (Figure 9.23). Like the Performance Profiler,

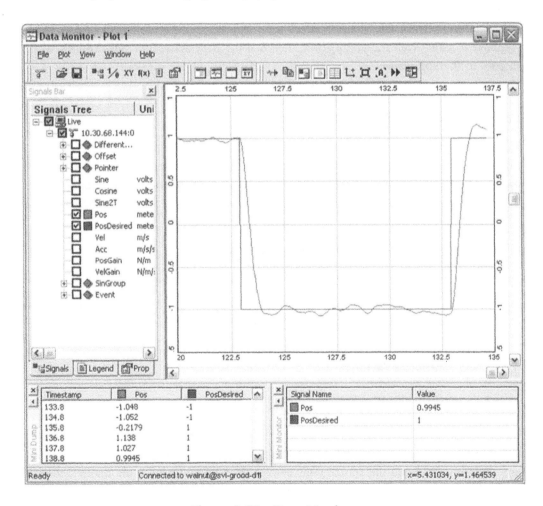

Figure 9.23: Data Monitor.

an agent on the target periodically samples the variables being monitored and stores the values in a local buffer. A low priority process then sends this buffer to the host for display. The performance hit is claimed to be fairly minor.

The sampling interval is configurable. Variables can be added or removed from the monitor list while the application runs. Variable values can also be modified at run time. Collected data can be stored to a file on the host, formatted for post processing by other applications, such as MatLab or Excel.

As described thus far, the Data Monitor is asynchronous with respect to any running applications. In some situations, it may be more meaningful to collect data synchronously. The Data Monitor target agent can be configured to collect data in response to calls from the application, thus making data collection synchronous with respect to the application.

Code Coverage Analyzer (Formerly CoverageScope)

This tool reports how much of the code in a system has been executed, or more importantly, which code hasn't been executed. Unlike the other tools, Code Coverage requires that the code be recompiled with the appropriate instrumentation added. You select which files to instrument and the level of coverage analysis. The four types of coverage are:

- **Function**: Verifies that the function was called.

- **Block:** Did this statement or block of statements get executed?

- **Decision:** Have both the true and false branches of a Boolean expression used in a branching statement such as if() or while() been executed?

- **Condition:** Did every subexpression in a Boolean expression evaluate to both true and false? Example:

```
if (a && b || BuggyFunction())
```

How to be sure that `BuggyFunction()` was executed?

The high-level output from Code Coverage is shown in Figure 9.24. Double-click on any function name and the source code shows up in an editor with the uncovered code highlighted.

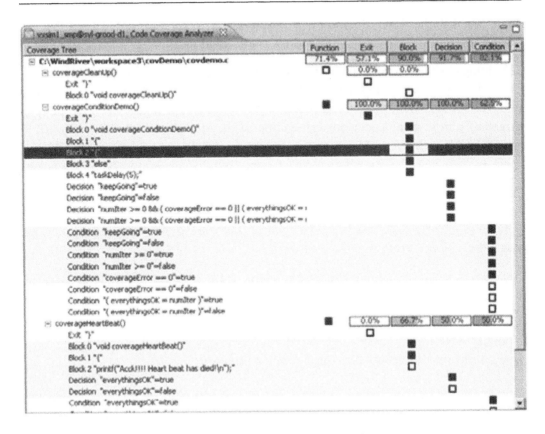

Figure 9.24: Code Coverage Analyzer.

Summary

In this chapter we've seen how some of the major players in the open source software business have adapted the Eclipse platform to create high quality commercial software development tools. Each vendor has chosen a slightly different approach that emphasizes its particular strengths to differentiate its offerings from the competition.

Free software doesn't necessarily mean free of charge, and in fact there's no such thing as zero cost software[2]. One way or another you're going to pay for it. You can download it all from the Internet and go through the inevitable learning

---

[2] As one open source practitioner put it some time ago, "Think free speech, not free beer."

curve. On the other hand, the most cost-effective way to get your product out the door on time may be to buy a high-quality commercial tool.

This brings us to the end of our exploration of Eclipse as a platform for developing embedded software around Linux. We've seen how Eclipse provides intuitive, graphical tools for building and managing software projects of any size. Even though Eclipse was originally aimed at Java development, additional plug-ins provide facilities to work with the GNU compiler tool chains for C and C++ development. For the embedded space in particular, Eclipse offers tools to access and manage remote target hardware.

We looked at the Eclipse plug-in architecture and how it extends the functionality of the basic platform. Well over a thousand plug-ins, both open source and commercial, provide support for just about any development task you require. If you can't find what you need, you can always create your own.

Needless to say, there's a lot more there. We've really only scratched the surface. I hope I've piqued your interest sufficiently to dive in and play around with it some more. And don't forget that Eclipse is constantly evolving.

You may even want to get involved in Eclipse development itself. The Eclipse community enthusiastically welcomes new contributors.

The Eclipse Public License

Open Source Initiative OSI - Eclipse Public License v 1.0

THE ACCOMPANYING PROGRAM IS PROVIDED UNDER THE TERMS OF THIS ECLIPSE PUBLIC LICENSE ("AGREEMENT"). ANY USE, REPRODUCTION OR DISTRIBUTION OF THE PROGRAM CONSTITUTES RECIPIENT'S ACCEPTANCE OF THIS AGREEMENT.

1. DEFINITIONS

"Contribution" means:

  a) in the case of the initial Contributor, the initial code and documentation distributed under this Agreement, and

  b) in the case of each subsequent Contributor:

   I.   changes to the Program, and

   II. additions to the Program;

where such changes and/or additions to the Program originate from and are distributed by that particular Contributor. A Contribution 'originates' from a Contributor if it was added to the Program by such Contributor itself or anyone acting on such Contributor's behalf. Contributions do not include additions to the Program which: (i) are separate modules of software distributed in conjunction with the Program under their own license agreement, and (ii) are not derivative works of the Program.

"Contributor" means any person or entity that distributes the Program.

"Licensed Patents" mean patent claims licensable by a Contributor which are necessarily infringed by the use or sale of its Contribution alone or when combined with the Program.

"Program" means the Contributions distributed in accordance with this Agreement.

"Recipient" means anyone who receives the Program under this Agreement, including all Contributors.

2.   GRANT OF RIGHTS

a)   Subject to the terms of this Agreement, each Contributor hereby grants Recipient a non-exclusive, worldwide, royalty-free copyright license to reproduce, prepare derivative works of, publicly display, publicly perform, distribute and sublicense the Contribution of such Contributor, if any, and such derivative works, in source code and object code form.

b)   Subject to the terms of this Agreement, each Contributor hereby grants Recipient a non-exclusive, worldwide, royalty-free patent license under Licensed Patents to make, use, sell, offer to sell, import and otherwise transfer the Contribution of such Contributor, if any, in source code and object code form. This patent license shall apply to the combination of the Contribution and the Program if, at the time the Contribution is added by the Contributor, such addition of the Contribution causes such combination to be covered by the Licensed Patents. The patent license shall not apply to any other combinations which include the Contribution. No hardware per se is licensed hereunder.

c)   Recipient understands that although each Contributor grants the licenses to its Contributions set forth herein, no assurances are provided by any Contributor that the Program does not infringe the patent or other intellectual property rights of any other entity. Each Contributor disclaims any liability to Recipient for claims brought by any other entity based on infringement of intellectual property rights or otherwise. As a condition to exercising the rights and licenses granted hereunder, each Recipient hereby assumes sole responsibility to secure any other intellectual property rights needed, if any. For example, if a third party patent license is required to allow Recipient to distribute the Program, it is Recipient's responsibility to acquire that license before distributing the Program.

d) Each Contributor represents that to its knowledge it has sufficient copyright rights in its Contribution, if any, to grant the copyright license set forth in this Agreement.

3. REQUIREMENTS

A Contributor may choose to distribute the Program in object code form under its own license agreement, provided that:

a) it complies with the terms and conditions of this Agreement; and

b) its license agreement:

    i) effectively disclaims on behalf of all Contributors all warranties and conditions, express and implied, including warranties or conditions of title and non-infringement, and implied warranties or conditions of merchantability and fitness for a particular purpose;

    ii) effectively excludes on behalf of all Contributors all liability for damages, including direct, indirect, special, incidental and consequential damages, such as lost profits;

    iii) states that any provisions which differ from this Agreement are offered by that Contributor alone and not by any other party; and

    iv) states that source code for the Program is available from such Contributor, and informs licensees how to obtain it in a reasonable manner on or through a medium customarily used for software exchange.

When the Program is made available in source code form:

a) it must be made available under this Agreement; and

b) a copy of this Agreement must be included with each copy of the Program.

Contributors may not remove or alter any copyright notices contained within the Program.

Each Contributor must identify itself as the originator of its Contribution, if any, in a manner that reasonably allows subsequent Recipients to identify the originator of the Contribution.

4.   COMMERCIAL DISTRIBUTION

Commercial distributors of software may accept certain responsibilities with respect to end users, business partners and the like. While this license is intended to facilitate the commercial use of the Program, the Contributor who includes the Program in a commercial product offering should do so in a manner which does not create potential liability for other Contributors. Therefore, if a Contributor includes the Program in a commercial product offering, such Contributor ("Commercial Contributor") hereby agrees to defend and indemnify every other Contributor ("Indemnified Contributor") against any losses, damages and costs (collectively "Losses") arising from claims, lawsuits and other legal actions brought by a third party against the Indemnified Contributor to the extent caused by the acts or omissions of such Commercial Contributor in connection with its distribution of the Program in a commercial product offering. The obligations in this section do not apply to any claims or Losses relating to any actual or alleged intellectual property infringement. In order to qualify, an Indemnified Contributor must: a) promptly notify the Commercial Contributor in writing of such claim, and b) allow the Commercial Contributor to control, and cooperate with the Commercial Contributor in, the defense and any related settlement negotiations. The Indemnified Contributor may participate in any such claim at its own expense.

For example, a Contributor might include the Program in a commercial product offering, Product X. That Contributor is then a Commercial Contributor. If that Commercial Contributor then makes performance claims, or offers warranties related to Product X, those performance claims and warranties are such Commercial Contributor's responsibility alone. Under this section, the Commercial Contributor would have to defend claims against the other Contributors related to those performance claims and warranties, and if a court requires any other Contributor to pay any damages as a result, the Commercial Contributor must pay those damages.

5.   NO WARRANTY

EXCEPT AS EXPRESSLY SET FORTH IN THIS AGREEMENT, THE PROGRAM IS PROVIDED ON AN "AS IS" BASIS, WITHOUT WARRANTIES OR CONDITIONS OF ANY KIND, EITHER EXPRESS OR IMPLIED INCLUDING, WITHOUT LIMITATION, ANY WARRANTIES OR CONDITIONS OF TITLE, NON-INFRINGEMENT, MERCHANTABILITY OR FITNESS FOR A PARTICULAR PURPOSE. Each Recipient is solely responsible for determining the appropriateness of using and distributing the Program and assumes all risks associated with its exercise of rights under this Agreement, including but not limited to the risks and costs of program errors, compliance with applicable laws, damage to or loss of data, programs or equipment, and unavailability or interruption of operations.

6. DISCLAIMER OF LIABILITY

EXCEPT AS EXPRESSLY SET FORTH IN THIS AGREEMENT, NEITHER RECIPIENT NOR ANY CONTRIBUTORS SHALL HAVE ANY LIABILITY FOR ANY DIRECT, INDIRECT, INCIDENTAL, SPECIAL, EXEMPLARY, OR CONSEQUENTIAL DAMAGES (INCLUDING WITHOUT LIMITATION LOST PROFITS), HOWEVER CAUSED AND ON ANY THEORY OF LIABILITY, WHETHER IN CONTRACT, STRICT LIABILITY, OR TORT (INCLUDING NEGLIGENCE OR OTHERWISE) ARISING IN ANY WAY OUT OF THE USE OR DISTRIBUTION OF THE PROGRAM OR THE EXERCISE OF ANY RIGHTS GRANTED HEREUNDER, EVEN IF ADVISED OF THE POSSIBILITY OF SUCH DAMAGES.

7. GENERAL

If any provision of this Agreement is invalid or unenforceable under applicable law, it shall not affect the validity or enforceability of the remainder of the terms of this Agreement, and without further action by the parties hereto, such provision shall be reformed to the minimum extent necessary to make such provision valid and enforceable.

If Recipient institutes patent litigation against any entity (including a cross-claim or counterclaim in a lawsuit) alleging that the Program itself (excluding combinations of the Program with other software or hardware) infringes such Recipient's patent(s), then such Recipient's rights granted under Section 2(b) shall terminate as of the date such litigation is filed.

All Recipient's rights under this Agreement shall terminate if it fails to comply with any of the material terms or conditions of this Agreement and does not cure such failure in a reasonable period of time after becoming aware of such noncompliance. If all Recipient's rights under this Agreement terminate, Recipient agrees to cease use and distribution of the Program as soon as reasonably practicable. However, Recipient's obligations under this Agreement and any licenses granted by Recipient relating to the Program shall continue and survive.

Everyone is permitted to copy and distribute copies of this Agreement, but in order to avoid inconsistency the Agreement is copyrighted and may only be modified in the following manner. The Agreement Steward reserves the right to publish new versions (including revisions) of this Agreement from time to time. No one other than the Agreement Steward has the right to modify this Agreement. The Eclipse Foundation is the initial Agreement Steward. The Eclipse Foundation may assign the responsibility to serve as the Agreement Steward to a suitable separate entity. Each new version of the Agreement will be given a distinguishing version number. The Program (including Contributions) may always be distributed subject to the version of the Agreement under which it was received. In addition, after a new version of the Agreement is published, Contributor may elect to distribute the Program (including its Contributions) under the new version. Except as expressly stated in Sections 2(a) and 2(b) above, Recipient receives no rights or licenses to the intellectual property of any Contributor under this Agreement, whether expressly, by implication, estoppel or otherwise. All rights in the Program not expressly granted under this Agreement are reserved.

This Agreement is governed by the laws of the State of New York and the intellectual property laws of the United States of America. No party to this Agreement will bring a legal action under this Agreement more than one year after the cause of action arose. Each party waives its rights to a jury trial in any resulting litigation.

The Embedded Linux Learning Kit

Chapter 6, "Device Software Development Platform," goes through the process of connecting Eclipse to a target computer for running and debugging applications. If you don't have a suitable target board, you might consider the Embedded Linux Learning Kit (E.L.L.K.) from Intellimetrix.

More than just a target board, the Embedded Linux Learning Kit is designed to teach embedded Linux in a practical, self-paced, hands-on environment, taking you step-by-step through the process of building and testing real embedded applications on real hardware. You'll learn how to:

- Set up boot parameters and boot Linux

- Configure and build the Linux kernel

- Build and debug application code over the network

- Access peripheral devices with and without device drivers

- Create network-based applications including a simple web server

The Embedded Linux Learning Kit uses a single board computer (SBC) with an ARM9 processor and a preloaded Linux kernel. Features include:

- 180 MHz, 200 MIPS ARM9 processor (Atmel AT91RM9200)

- 64 MB of DRAM

- 8 MB SPI serial Flash

- 256 MB NAND Flash

- 16 KB bootloader EEPROM

- SD/MMC socket

- Parallel LCD interface

- 10/100 Ethernet port

- USB 2.0 host and device ports

- IrDA transceiver

- RS-232 serial port for debugging

- On-board temperature sensor

- User-programmable LEDs and switches

- Optional LCD graphics display

The complete kit includes:

- Single board computer

- Power supply

- Serial and Ethernet crossover cables

- Pre-installed Linux kernel and U-Boot boot loader

- CD with support software:
    - Cross-development tools
    - Kernel source
    - Eclipse IDE
    - Tutorial code samples

- User's guide

To learn more, visit http://www.intellimetrix.us/. A discounted price is available to owners of this book. Go to www.intellimetrix.us/eclipsekit.htm to place an order at the reduced price.

Index

Printed and bound by CPI Group (UK) Ltd, Croydon, CR0 4YY

03/10/2024

01040336-0001